Nathan Jacobson

Lectures in Abstract Algebra

I. Basic Concepts

Springer-Verlag   New York Heidelberg Berlin

Nathan Jacobson
Department of Mathematics
Yale University
New Haven, Connecticut 06520

Managing Editor

P. R. Halmos
Indiana University
Department of Mathematics
Swain Hall East
Bloomington, Indiana 47401

Editors

F. W. Gehring
University of Michigan
Department of Mathematics
Ann Arbor, Michigan 48104

C. C. Moore
University of California at Berkeley
Department of Mathematics
Berkeley, California 94720

AMS Subject Classifications
06-01, 12-01, 13-01

Library of Congress Cataloging in Publication Data

Jacobson, Nathan, 1910–
    Lectures in abstract algebra.

    (Graduate texts in mathematics; 30-32)
    Reprint of the 1951–1964 ed. published by Van Nostrand, New York in The University series in higher mathematics.
    Bibliography: v. 3, p.
    Includes indexes.
    CONTENTS: 1. Basic concepts. 2. Linear algebra. 3. Theory of fields and Galois theory.
    1. Algebra, Abstract. I. Title. II. Series.
QA162.J3   1975   512′.02   75–15564

Softcover reprint of the hardcover 1st edition 1951

Originally published in the University Series in Higher Mathematics (D. Van Nostrand Company); edited by M. H. Stone, L. Nirenberg and S. S. Chern.

ISBN-13: 978-1-4684-7303-2   e-ISBN-13: 978-1-4684-7301-8
DOI: 10.1007/978-1-4684-7301-8

TO
MY WIFE

PREFACE

The present volume is the first of three that will be published under the general title *Lectures in Abstract Algebra*. These volumes are based on lectures which the author has given during the past ten years at the University of North Carolina, at The Johns Hopkins University, and at Yale University. The general plan of the work is as follows: The present first volume gives an introduction to abstract algebra and gives an account of most of the important algebraic concepts. In a treatment of this type it is impossible to give a comprehensive account of the topics which are introduced. Nevertheless we have tried to go beyond the foundations and elementary properties of the algebraic systems. This has necessitated a certain amount of selection and omission. We feel that even at the present stage a deeper understanding of a few topics is to be preferred to a superficial understanding of many.

The second and third volumes of this work will be more specialized in nature and will attempt to give comprehensive accounts of the topics which they treat. Volume II will bear the title *Linear Algebra* and will deal with the theory of vector spaces. Volume III, *The Theory of Fields and Galois Theory*, will be concerned with the algebraic structure of fields and with valuations of fields.

All three volumes have been planned as texts for courses. A great many exercises of varying degrees of difficulty have been included. Some of these perhaps rate stars, but we have felt that the disadvantages of the system of starring difficult exercises outweigh its advantages. A few sections have been starred (notation: *1) to indicate that these can be omitted without jeopardizing the understanding of subsequent material.

We are indebted to a great many friends for helpful criticisms and encouragement during the course of preparation of this volume. Professors A. H. Clifford, G. Hochschild and R. E. Johnson, Drs. D. T. Finkbeiner and W. H. Mills have read parts of the manuscript and given us useful suggestions for improving it. Drs. Finkbeiner and Mills have assisted with the proofreading. I take this opportunity to offer my sincere thanks to all of these men.

N. J.

New Haven, Conn.
January 22, 1951

CONTENTS

INTRODUCTION: CONCEPTS FROM SET THEORY
THE SYSTEM OF NATURAL NUMBERS

SECTION	PAGE
1. Operations on sets	2
2. Product sets, mappings	3
3. Equivalence relations	4
4. The natural numbers	7
5. The system of integers	10
6. The division process in I	12

CHAPTER I: SEMI-GROUPS AND GROUPS

1. Definition and examples of semi-groups	15
2. Non-associative binary compositions	18
3. Generalized associative law. Powers	20
4. Commutativity	21
5. Identities and inverses	22
6. Definition and examples of groups	23
7. Subgroups	24
8. Isomorphism	26
9. Transformation groups	27
10. Realization of a group as a transformation group	28
11. Cyclic groups. Order of an element	30
12. Elementary properties of permutations	34
13. Coset decompositions of a group	37
14. Invariant subgroups and factor groups	40
15. Homomorphism of groups	41
16. The fundamental theorem of homomorphism for groups	43
17. Endomorphisms, automorphisms, center of a group	45
18. Conjugate classes	47

CHAPTER II: RINGS, INTEGRAL DOMAINS AND FIELDS

SECTION   PAGE

1. Definition and examples 49
2. Types of rings 53
3. Quasi-regularity. The circle composition 55
4. Matrix rings 56
5. Quaternions 60
6. Subrings generated by a set of elements. Center 63
7. Ideals, difference rings 64
8. Ideals and difference rings for the ring of integers 66
9. Homomorphism of rings 68
10. Anti-isomorphism 71
11. Structure of the additive group of a ring. The charateristic
     of a ring 74
12. Algebra of subgroups of the additive group of a ring. One-
     sided ideals 75
13. The ring of endomorphisms of a commutative group 78
14. The multiplications of a ring 82

CHAPTER III: EXTENSIONS OF RINGS AND FIELDS

1. Imbedding of a ring in a ring with an identity 84
2. Field of fractions of a commutative integral domain 87
3. Uniqueness of the field of fractions 91
4. Polynomial rings 92
5. Structure of polynomial rings 96
6. Properties of the ring $\mathfrak{A}[x]$ 97
7. Simple extensions of a field 100
8. Structure of any field 103
9. The number of roots of a polynomial in a field 104
10. Polynomials in several elements 105
11. Symmetric polynomials 107
12. Rings of functions 110

CHAPTER IV: ELEMENTARY FACTORIZATION THEORY

1. Factors, associates, irreducible elements 114
2. Gaussian semi-groups 115
3. Greatest common divisors 118
4. Principal ideal domains 121

SECTION   PAGE

5. Euclidean domains 122
6. Polynomial extensions of Gaussian domains 124

CHAPTER V: GROUPS WITH OPERATORS

1. Definition and examples of groups with operators 128
2. M-subgroups, M-factor groups and M-homomorphisms . . . 130
3. The fundamental theorem of homomorphism for M-groups . 132
4. The correspondence between M-subgroups determined by a homomorphism 133
5. The isomorphism theorems for M-groups 135
6. Schreier's theorem 137
7. Simple groups and the Jordan-Hölder theorem 139
8. The chain conditions 142
9. Direct products 144
10. Direct products of subgroups 145
11. Projections 149
12. Decomposition into indecomposable groups 152
13. The Krull-Schmidt theorem 154
14. Infinite direct products 159

CHAPTER VI: MODULES AND IDEALS

1. Definitions 162
2. Fundamental concepts 164
3. Generators. Unitary modules 166
4. The chain conditions 168
5. The Hilbert basis theorem 170
6. Noetherian rings. Prime and primary ideals 172
7. Representation of an ideal as intersection of primary ideals 175
8. Uniqueness theorems 177
9. Integral dependence 181
10. Integers of quadratic fields 184

CHAPTER VII: LATTICES

1. Partially ordered sets 187
2. Lattices 189
3. Modular lattices 193
4. Schreier's theorem. The chain conditions 197

SECTION   PAGE

5. Decomposition theory for lattices with ascending chain condition 201

6. Independence 202

7. Complemented modular lattices 205

8. Boolean algebras 207

    Index 213

Introduction

CONCEPTS FROM SET THEORY
THE SYSTEM OF NATURAL NUMBERS

The purpose of this volume is to give an introduction to the basic algebraic systems: groups, rings, fields, groups with operators, modules, and lattices. The study of these systems encompasses a major portion of classical algebra. Thus, in a sense our subject matter is old. However, the axiomatic development which we have adopted here is comparatively new. A beginner may find our account at times uncomfortably abstract since we do not tie ourselves down to the study of one particular system (e.g., the system of real numbers). Supplementary study of the exercises and examples should help to overcome this difficulty. At any rate, it will be obvious that much time is saved and a clearer insight is eventually achieved by the present method.

The basic ingredients of the systems that we shall study are sets and mappings of these sets. Notions from set theory will occur constantly in our discussion. Hence, it will be useful to consider briefly in the first part of this Introduction some of these ideas before embarking on the study of the algebraic systems. We shall not attempt to be completely rigorous in our sketchy account of the elements of set theory. The reader should consult the standard texts for systematic and detailed accounts of this subject. Of these we single out Bourbaki's *Théorie des Ensembles* as particularly appropriate for our purposes.

The second part of this Introduction sketches a treatment of the system P of natural numbers as an abstract mathematical system. The starting point here is a set and a mapping in the set (the successor mapping) that is assumed to satisfy Peano's axioms. By means of this, one can introduce addition, multiplication, and the relation of order in P. We shall also define the system I of integers as a certain extension of the system P of natural numbers. Finally, we shall derive one or two arithmetic facts concerning I that are indispensable in elementary group theory. Full accounts of the foundations of the system of natural numbers are available in Landau's *Grundlagen der Analysis* and in Graves' *Theory of Functions of Real Variables*.

1. Operations on sets. We begin our discussion with a brief survey of the fundamental concepts of the theory of sets.

Let S be an arbitrary set (or collection) of elements $a, b, c, \cdots$. The nature of the elements is immaterial to us. We indicate the fact that an element a is in S by writing $a \,\varepsilon\, S$ or $S \ni a$. If A and B are two subsets of S, then we say that A is *contained* in B or B *contains* A (notation: $A \subseteq B$ or $B \supseteq A$) if every a in A is also in B. The statement $A = B$ thus means that $A \supseteq B$ and $B \supseteq A$. Also we write $A \supset B$ if $A \supseteq B$ but $B \neq A$. In this case A is said to contain B properly, or B is a *proper subset* of A.

If A and B are any two subsets of S, the collection of elements c such that $c \,\varepsilon\, A$ and $c \,\varepsilon\, B$ is called the *intersection* $A \cap B$ of A and B. More generally we can define the intersection of any finite number of sets, and still more generally, if $\{A\}$ denotes any collection of subsets of S, then we define the intersection $\cap A$ as the set of elements c such that $c \,\varepsilon\, A$ for every A in $\{A\}$. If the collection $\{A\}$ is finite, so that its members can be denoted as $A_1, A_2, \cdots, A_n$, then the intersection can be written as $\bigcap_1^n A_i$ or as $A_1 \cap A_2 \cap \cdots \cap A_n$.

Similar remarks apply to logical sums of subsets of S. The *logical sum* or *union* of the collection $\{A\}$ of subsets A is the set of elements u such that $u \,\varepsilon\, A$ for at least one A in $\{A\}$. We denote this set as $\cup A$ or, if the collection is finite, as $\bigcup_1^n A_i$ or $A_1 \cup A_2 \cup \cdots \cup A_n$.

The collection of all subsets of the given set S will be denoted as $P(S)$. In order to avoid considering exceptional cases it is necessary to count the whole set S and the vacuous set as members of $P(S)$. One may regard the latter as a zero element that is adjoined to the collection of "real" subsets. We use the notation $\varnothing$ for the vacuous set. The convenience of introducing this set is illustrated in the use of the equation $A \cap B = \varnothing$ to indicate that A and B are non-overlapping, that is, they have no elements in common. If S is a finite set of n elements, then $P(S)$ consists of $\varnothing$, n sets containing single elements, $\cdots$, $\binom{n}{i} = \dfrac{n(n-1)\cdots(n-i+1)}{1 \cdot 2 \cdots i}$ sets containing i elements, and so on. Hence the total number of elements in $P(S)$ is

$$1 + \binom{n}{1} + \binom{n}{2} + \cdots + \binom{n}{n} = (1+1)^n = 2^n.$$

2. Product sets, mappings. If S and T are arbitrary sets, we define the *product set* $S \times T$ to be the collection of pairs (s,t), s in S, t in T. The two sets S and T need not be distinct. In the product $S \times T$ the elements (s,t) and (s',t') are regarded as equal if and only if $s = s'$ and $t = t'$. Thus if S consists of the m elements $s_1, s_2, \cdots, s_m$ and T consists of the n elements $t_1, t_2, \cdots, t_n$, then $S \times T$ consists of the mn elements (s_i, t_j). More generally, if $S_1, S_2, \cdots, S_r$ are any sets, then ΠS_i or $S_1 \times S_2 \times \cdots \times S_r$ is defined to be the collection of r-tuples $(s_1, s_2, \cdots, s_r)$ where the ith component s_i is in the set S_i.

A (single-valued) *mapping* α of a set S *into* a set T is a correspondence that associates with each $s \in S$ a single element $t \in T$. It is customary in elementary mathematics to write the image in T of s as $\alpha(s)$. We shall find it more convenient to denote this element as $s\alpha$ or s^α. With the mapping α we can associate the subset of $S \times T$ consisting of the points $(s, s\alpha)$. We shall call this set the *graph* of α. Its characteristic properties are:

1. If s is any element of S, then there is an element of the form (s,t) in the graph.
2. If (s,t_1) and (s,t_2) are in the graph, then $t_1 = t_2$.

A mapping α is said to be a mapping of S *onto* T if every $t \in T$ occurs as an image of some $s \in S$. In any case we shall denote the image set (= set of image elements) of S under α as $S\alpha$ or S^α. A mapping α of S into T is said to be 1–1 if $s_1\alpha = s_2\alpha$ holds only if $s_1 = s_2$, that is, distinct points of S have distinct images. Suppose now that α is a 1–1 mapping of S onto T. Then if t is any element in T, there exists a unique element s in S such that $s\alpha = t$. Hence if we associate with t this element s we obtain a mapping of T into S. We shall call this mapping the *inverse mapping* α^{-1} of α. It is immediate that α^{-1} is 1–1 of T onto S.

It is natural to regard two mappings α and β of S into T as equal if and only if $s\alpha = s\beta$ for all s in S. This means that $\alpha = \beta$ if and only if these mappings have the same graph.

Let α be a mapping of S into T and let β be a mapping of T into a third set U. The mapping that sends the element s of S into the element $(s\alpha)\beta$ of U is called the *resultant* or *product* of α and β. We denote this mapping as $\alpha\beta$, so that by definition $s(\alpha\beta) = (s\alpha)\beta$.

Mappings of a set into itself will be called *transformations* of the set. Among these are included the *identity mapping* or *transformation* that leaves every element of S fixed. We denote this mapping as 1 (or 1_S if this is necessary). If α is any transformation of S, it is clear that $\alpha 1 = \alpha = 1\alpha$.

If α is a 1–1 mapping of S onto T and α^{-1} is its inverse, then $\alpha\alpha^{-1} = 1_S$ and $\alpha^{-1}\alpha = 1_T$. The following useful converse of this remark is also easy to verify: If α is a mapping of S into T, and β is a mapping of T into S such that $\alpha\beta = 1_S$ and $\beta\alpha = 1_T$, then α and β are 1–1, onto mappings and $\beta = \alpha^{-1}$.

The concept of a product set permits us to define the notion of a function of two or more variables. Thus a function of two variables in S with values in T is a mapping of $S \times S$ into T. More generally we can consider mappings of $S_1 \times S_2$ into T. Of particular interest for us will be the mappings of $S \times S$ into S. We shall call such mappings *binary compositions* in the set S.

3. Equivalence relations. We say that a *relation R* is defined in a set S if, for any ordered pair of elements (a,b), a,b in S, we can determine whether or not a is in the given relation to b. More precisely, a relation can be defined as a mapping of the set $S \times S$ into a set consisting of two elements. We can take these to be the words "yes" and "no." Then if $(a,b) \rightarrow$ yes (that is, is mapped into "yes"), we say that a is in the given relation to b.

In this case we write $a\,R\,b$. If $(a,b) \to$ no, then we say that a is not in the given relation to b and we write $a\,\not{R}\,b$.

A relation $\sim$ (in place of R) is called an *equivalence relation* if it satisfies the following conditions:

1. $a \sim a$ (reflexive property).
2. $a \sim b$ implies $b \sim a$ (symmetric property).
3. $a \sim b$ and $b \sim c$ imply that $a \sim c$ (transitive property).

An example of an equivalence relation is obtained by letting S be the collection of points in the plane and by defining $a \sim b$ if a and b lie on the same horizontal line. If $a \,\varepsilon\, S$, it is clear that the collection $\bar{a}$ of elements $b \sim a$ is the horizontal line through the point a. The collection of these lines gives a decomposition of the set S into non-overlapping subsets. We shall now show that this phenomenon is typical of equivalence relations.

Let S be any set and let $\sim$ be any equivalence relation in S. If $a \,\varepsilon\, S$, let $\bar{a}$ denote the subset of S of elements b such that $b \sim a$. By 1, $a \,\varepsilon\, \bar{a}$ and by 2 and 3, if b_1 and $b_2 \,\varepsilon\, \bar{a}$, then $b_1 \sim b_2$. Hence $\bar{a}$ is a collection of equivalent elements. Moreover, $\bar{a}$ is a maximal collection of this type; for, if c is any element equivalent to some b in $\bar{a}$, then $c \,\varepsilon\, \bar{a}$. We call $\bar{a}$ the *equivalence class* determined by (or containing) the element a. If $b \,\varepsilon\, \bar{a}$, then $\bar{b} \subseteq \bar{a}$; hence by the maximality of $\bar{b}$, $\bar{b} = \bar{a}$. This implies the important conclusion that any two equivalence classes are either identical or they have a vacuous intersection. Hence, the collection of distinct equivalence classes gives a decomposition of the set S into non-intersecting sets.

Conversely, suppose that a given set S is decomposed in any way into sets $A, B, \cdots$ no two of which overlap. Then we can define an equivalence relation in S by specifying that $a \sim b$ if the sets A, B containing a and b respectively are identical. It is clear that this relation has the required properties. Also, obviously, the equivalence classes determined by this relation are just the given sets $A, B, \cdots$.

The collection $\bar{S}$ of equivalence classes defined by an equivalence relation in S is called the *quotient set* of S relative to the given relation. It should be emphasized that $\bar{S}$ is not a subset of S but rather a subset of the collection $P(S)$ of subsets of S.

There is an intimate connection between equivalence relations and mappings. In the first place, if S is a set and $\bar{S}$ is its quotient set relative to an equivalence relation, then we have a natural mapping ν of S onto $\bar{S}$. This is defined by the rule that the element a of S is sent into the equivalence class $\bar{a}$ determined by a. Evidently this mapping is a mapping onto $\bar{S}$.

On the other hand, suppose that we are given any mapping α of the set S onto a second set T. Then we can use α to define an equivalence relation. Our rule here is that $a \sim b$ if $a\alpha = b\alpha$. Clearly this satisfies the axioms 1, 2 and 3. If a' is an element of T and a is an element of S such that $a\alpha = a'$, then the equivalence class $\bar{a}$ is just the set of elements of S that are mapped into a'. We call this set the inverse image of a' and we denote it as $\alpha^{-1}(a')$.

Suppose now that $\sim$ is any equivalence relation in S with quotient set $\bar{S}$. Let α be a mapping of S onto T which has the property that the inverse images $\alpha^{-1}(a')$ are logical sums of sets belonging to $\bar{S}$. This is equivalent to saying that any set belonging to $\bar{S}$ is contained in some inverse image $\alpha^{-1}(a')$. Hence it means simply that, if a and b are any two elements of S such that $a \sim b$, then $a\alpha = b\alpha$. It is therefore clear that the rule $\bar{a} \rightarrow a\alpha$ defines a mapping of $\bar{S}$ onto T. We denote this mapping as $\bar{\alpha}$ and call it the mapping of $\bar{S}$ *induced by* the given mapping α. The defining equation $\bar{a}\bar{\alpha} = a\alpha$ shows that the original mapping is the resultant of the natural mapping $a \rightarrow \bar{a}$ and the mapping $\bar{\alpha}$, that is, $\alpha = \nu\bar{\alpha}$.

This type of factorization of mappings will play an important role in the sequel. It is particularly useful when the set of inverse images $\alpha^{-1}(a')$ coincides with $\bar{S}$; for, in this case, the mapping $\bar{\alpha}$ is 1–1. Thus if $\bar{a}\bar{\alpha} = \bar{b}\bar{\alpha}$, then $a\alpha = b\alpha$ and $a \sim b$. Hence $\bar{a} = \bar{b}$. Thus we obtain here a factorization $\alpha = \nu\bar{\alpha}$ where $\bar{\alpha}$ is 1–1 onto T and ν is the natural mapping.

As an illustration of our discussion we consider the perpendicular projection π_x of the plane S onto the x-axis T. Here a point a is sent into the foot of the perpendicular joining it to the x-axis. If a' is a point on the x-axis, $\pi_x^{-1}(a')$ is the set of points on the vertical line through a'. The set of inverse images is the collection of these vertical lines, and the induced mapping $\bar{\pi}_x$

sends a vertical line into its intersection with the x-axis. Clearly this mapping is 1–1, and $\pi_x = \nu\bar{\pi}_x$ where ν is the natural mapping of a point into the vertical line containing it.

4. The natural numbers. The system of natural numbers 1, 2, 3, $\cdots$ is fundamental in algebra in two respects. In the first place, it serves as a starting point for constructing examples of more elaborate systems. Thus we shall use this system to construct the system of integers, the system of rational numbers, of residue classes modulo an integer, etc. In the second place, in studying algebraic systems, functions or mappings of the set of natural numbers play an important role. For example, in a system in which an associative multiplication is defined, the powers a^n of a fixed a determine a function or mapping $n \to a^n$ of the set of natural numbers.

We shall begin with the following assumptions (essentially Peano's axioms) concerning the set P of natural numbers.

1. P is not vacuous.

2. There exists a 1–1 mapping $a \to a^+$ of P into itself. (a^+ is the immediate successor of a.)

3. The set of images under the successor mapping is a proper subset of P.

4. Any subset of P that contains an element that is not a successor and that contains the successor of every element in the set coincides with P. This is called the *axiom of induction*.

All the properties that we shall state concerning P are consequences of these axioms. By 3 and 4 any two elements of P that are not successors are equal. As usual, we denote the unique non-successor as 1. Also we set $1^+ = 2, 2^+ = 3$, etc.

Property 4 is the basis of proofs by the *first principle of induction*. This can be stated as follows: Suppose that for each natural number n there is associated a statement $E(n)$. Suppose that $E(1)$ is true and that $E(r^+)$ is true whenever $E(r)$ is true. Then $E(n)$ is true for all n. This follows directly from 4. Thus let S be the set of natural numbers s for which $E(s)$ is true. This set contains 1 and it contains r^+ for every $r \, \varepsilon \, S$. Hence $S = P$ and this means that $E(n)$ is true for all n in P.

EXERCISE

1. Prove that $n^+ \neq n$ for every n.

Addition of natural numbers is defined to be a binary composition in P such that the value $x + y$ for the pair x,v satisfies

(a) $$1 + y = y^+$$

(b) $$x^+ + y = (x + y)^+.$$

It can be shown that such a function exists and is unique. Moreover, one has the following basic properties:

A1 $x + (y + z) = (x + y) + z$ (associative law)

A2 $x + y = y + x$ (commutative law)

A3 $x + z = y + z$ implies that $x = y$ (cancellation law).

The proofs of these results and the ones on multiplication and order that follow will be omitted. These can be found in the above-mentioned texts.

Multiplication in P is a binary composition satisfying

(a) $$1y = y$$

(b) $$x^+y = xy + y.$$

Such a composition exists, is unique, and has the usual properties:

M1 $$x(yz) = (xy)z$$

M2 $$xy = yx$$

M3 $$xz = yz \quad \text{implies that} \quad x = y.$$

Also we have the following fundamental rule connecting addition and multiplication

D $$x(y + z) = xy + xz \quad \text{(distributive law)}.$$

The third fundamental concept in the system P is that of *order*. This can be defined in terms of addition by stating that a is greater than b ($a > b$ or $b < a$) if the equation $a = b + x$

has a solution for x in P. The following are the basic properties of this relation:

O1 $x > y$ excludes $x \leq y$ (asymmetry)

O2 $x > y$ and $y > z$ imply $x > z$ (transitivity)

O3   For any ordered pair (x,y) one and only one of the following holds: $x > y$, $x = y$, $x < y$ (trichotomy). (Note that this implies O1. We include both of these since one is often interested in systems in which O1 and O2 hold but not O3.)

O4   In any non-vacuous set of natural numbers there is a least number, that is, a number l of the set such that $l \leq s$ for all s in the set.

Proof of O4.   Let S be the given set and M the set of natural numbers m that satisfy $m \leq s$ for every $s \, \varepsilon \, S$. 1 is in M. If s is a particular element in S, then $s^+ > s$ and hence $s^+ \notin M$. Hence $M \neq P$. By the principle of induction there exists a natural number l such that $l \, \varepsilon \, M$ but $l^+ \notin M$. Then l is the required number; for $l \leq s$ for every s and $l \, \varepsilon \, S$ since otherwise $l < s$ for every s in S. Then $l^+ \leq s$ contradicting $l^+ \notin M$.

The property O4 is called the *well-ordering* property of P. It is the basis of the following *second principle of induction*. Suppose that for each $n \, \varepsilon \, P$ we have a statement $E(n)$. Suppose that it is known that $E(r)$ is true for a particular r if $E(s)$ is true for all $s < r$. (This implies that it is known that $E(1)$ is true.) Then $E(n)$ is true for all n. To prove this let F be the set of elements r such that $E(r)$ is not true. If F is not vacuous, let t be its least element. Then $E(t)$ is not true but $E(s)$ is true for all $s < t$. This contradicts our assumption. Hence F is vacuous and $E(n)$ is true for all n.

The main relations between order and addition, and order and multiplication are given in the following statements:

OA $a > b$ implies and is implied by $a + c > b + c$.

OM $a > b$ implies and is implied by $ac > bc$.

EXERCISE

1. Prove that if $a > b$ and $c > d$, then $a + c > b + d$ and $ac > bd$.

5. The system of integers.   Instead of following the usual procedure of adding to the system P a 0 element and the negatives we shall obtain the extended system in a way that seems more natural and intuitive.   We shall construct a new system I of integers that contains a subsystem which is essentially the same as the set of natural numbers.

We consider first the set $P \times P$ of ordered pairs of natural numbers (a,b).   In this set we introduce the relation $(a,b) \sim (c,d)$ if $a + d = b + c$.   It is easy to verify that this is an equivalence relation.   What we have in mind, of course, in making this definition is that the equivalence class $\overline{(a,b)}$ determined by (a,b) is to play the role of the difference of a and b.   If we represent the pair (a,b) in the usual way as the point with abscissa a and ordinate b, then $\overline{(a,b)}$ is the set of points with natural number coordinates on the line of slope 1 through (a,b).   We call the equivalence

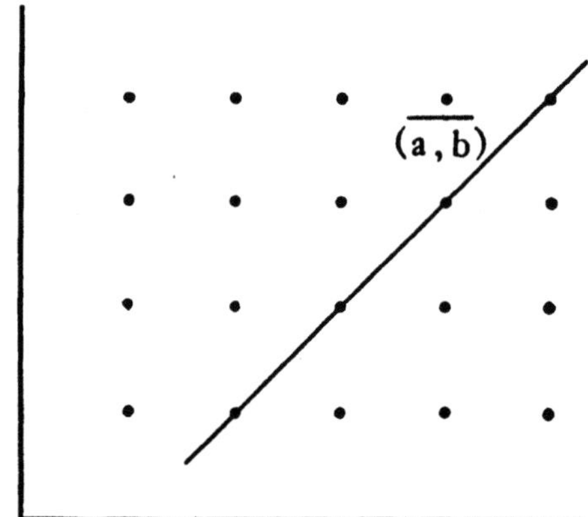

classes $\overline{(a,b)}$ *integers* and we denote their totality as I.   As a preliminary to defining addition we note that, if $(a,b) \sim (a',b')$ and $(c,d) \sim (c',d')$, then $(a + c, b + d) \sim (a' + c', b' + d')$; for the hypotheses are that $a + b' = a' + b$ and $c + d' = c' + d$. Hence $a + c + b' + d' = a' + c' + b + d$, which means that $(a + c, b + d) \sim (a' + c', b' + d')$.   It follows that the integer

$(\overline{a + c, \, b + d})$ is a function of $(\overline{a,b})$ and $(\overline{c,d})$. We define this integer to be the *sum* of the integers $(\overline{a,b})$ and $(\overline{c,d})$:

$$(\overline{a,b}) + (\overline{c,d}) = (\overline{a + c, \, b + d}).$$

It is easy to verify that the rules A1, A2, A3 hold. Also we note that $(a,a) \sim (b,b)$ and if we set $0 = (\overline{a,a})$, then

A4 $\qquad\qquad 0 + x = x \quad$ for every x in I.

Finally every integer has a negative: If $x = (\overline{a,b})$, then we denote $(\overline{b,a})$ as $-x$ and we have

A5 $\qquad\qquad\qquad x + (-x) = 0.$

We note next that, if $(a,b) \sim (a',b')$ and $(c,d) \sim (c',d')$, then $a + b' = a' + b,\, c + d' = c' + d$. Hence

$$c(a + b') + d(a' + b) + a'(c + d') + b'(c' + d)$$
$$= c(a' + b) + d(a + b') + a'(c' + d) + b'(c + d')$$

so that

$$ac + b'c + a'd + bd + a'c + a'd' + b'c' + b'd$$
$$= a'c + bc + ad + b'd + a'c' + a'd + b'c + b'd'.$$

The cancellation law gives

$$ac + bd + a'd' + b'c' = bc + ad + a'c' + b'd'.$$

This shows that $(ac + bd,\, ad + bc) \sim (a'c' + b'd',\, a'd' + b'c')$. Hence, if we define

$$(\overline{a,b})(\overline{c,d}) = (\overline{ac + bd,\, ad + bc}),$$

we obtain a single-valued function. It can be verified that this product function is associative and commutative and distributive with respect to addition. The cancellation law holds if the factor z to be cancelled is not 0.

We regard the integer $(\overline{a,b}) > (\overline{c,d})$ if $a + d > b + c$. This relation is well defined. One can verify easily that O1, O2, O3 and OA hold. The property OM has to be modified to state that

OM' $\quad$ If $z > 0$, then $x > y$ if and only if $xz > yz$.

EXERCISE

1. Show that, if $x > y$, then $-x < -y$.

We consider now the set P' of positive integers. By definition this set is the subset of I of elements $x > 0$. If $x = \overline{(a,b)}$, $x > 0$ is equivalent to the requirement that $a > b$. Hence $x = \overline{(b + u,b)}$ and it is immediate that $(b + u,b) \sim (c + u,c)$. Now let u be any natural number (element of P) and define u' to be the positive integer $\overline{(b + u,b)}$. Our remarks show that the mapping $u \to u'$ is a single-valued mapping of P onto P'. Moreover, if $(b + u,b) \sim (c + v,c)$, then $b + u + c = b + c + v$ so that $u = v$. Hence $u \to u'$ is 1–1. We leave it to the reader to verify the following properties of our correspondence:

$$(u + v)' = u' + v'$$

$$(uv)' = u'v'$$

$$u > v \quad \text{is equivalent to} \quad u' > v'.$$

Thus, we obtain the same result if (1) we add two natural numbers and then take the positive integer corresponding to the result, or (2) we add the positive integers corresponding to the natural numbers. A similar statement holds for multiplication. Because of this situation we can discard the original system of natural numbers and use in its place the system of positive integers. Also we can appropriate the notations originally used for P for the system of positive integers. Hence, from now on we denote the latter as P and we denote its numbers as 1, 2, 3, $\cdots$. The remaining numbers of I are then 0, -1, -2, $\cdots$.

EXERCISES

1. Prove that any non-vacuous set S of integers that is bounded below (above), in the sense that there exists an integer b (B) such that $b \leq s$ ($B \geq s$) for every s in S, has a least (greatest) element.

2. If $x \geq 0$, we set $|x| = x$ and, if $x < 0$, we set $|x| = -x$. Prove the rules $|xy| = |x||y|, |x + y| \leq |x| + |y|$.

6. The division process in I. We shall obtain some of the elementary arithmetic properties of I in the course of our discussion of groups and integral domains. The starting point in the study of the arithmetic of I is the following familiar result.

Theorem. *If a is any integer and $b \neq 0$, then there exist integers $q, r, 0 \leq r < |b|$, such that $a = bq + r$.*

Proof. Consider the multiples $x|b|$ of $|b|$ that are $\leq a$. The collection M of these multiples is not vacuous since $-|a||b| \leq -|a| \leq a$. Hence, the set M has a greatest member $h|b|$. Then $h|b| \leq a$ so that $a = h|b| + r$ where $r \geq 0$. On the other hand $(h+1)|b| = h|b| + |b| > h|b|$. Hence $(h+1)|b| > a$ and $h|b| + |b| > h|b| + r$. Thus, $r < |b|$. We now set $q = h$ if $b > 0$ and $q = -h$ if $b < 0$. Then $h|b| = qb$ and $a = qb + r$ as required.

EXERCISE

1. Prove that q and r are unique.

We shall say that the integer b is a *factor* or *divisor* of the integer a if there exists a $c \in I$ such that $a = bc$. Also a is called a *multiple* of b and we denote this relation by $b \mid a$. Clearly this is a transitive relation. If $b \mid a$ and $a \mid b$, we have $a = bc$ and $b = ad$. Hence, $a = adc$. If $a \neq 0$, the cancellation law implies that $dc = 1$. Hence, $|d||c| = 1$ and $d = \pm 1, c = \pm 1$. This shows that if $b \mid a$ and $a \mid b$ and $a \neq 0$, then $a = \pm b$.

An integer d is called a *greatest common divisor* (g.c.d.) of a and b if (1) $d \mid a$ and $d \mid b$ and (2) if e is any common factor of a and b, then $e \mid d$. The existence of a g.c.d. for any pair a,b with $a \neq 0$ is easily proved by using the division process given in the above theorem. For this purpose we consider the totality D of integers of the form $ax + by$. This set includes positive integers. Hence, there is a least positive integer $d = at + bs$ in the set. Now $a = dq + r$ where $0 \leq r < d$. Also $r = a - dq = a(1 - qt) + b(-qs) \in D$. Since d is the least positive integer in D, $r = 0$. Hence, $d \mid a$. Similarly $d \mid b$. Next let $e \mid a$ and $e \mid b$. Then $e \mid at$ and $e \mid bs$. Hence, $e \mid (at + bs)$. Thus $e \mid d$.

If d' is a second greatest common divisor of a and b, (2) implies that $d \mid d'$ and $d' \mid d$. Hence $d' = \pm d$. We have seen that we can always take d to be ≥ 0. This particular greatest common divisor will be denoted as (a,b).

The existence of greatest common divisors serves as a basis for the proof of the *fundamental theorem of arithmetic* that any positive integer can be written in one and only one way as a product of positive primes. By a *prime* p we mean an integer that is divisible only by p, $-p$, 1, -1. We shall obtain this result later (Chapter IV) in our study of arithmetic properties of integral domains. Also one can prove easily either by using the fundamental theorem or by using simple properties of greatest common divisors that the integer

$$m = ab/(a,b)$$

is a *least common multiple* of a and b. By this we mean that m is a multiple of a and b and any common multiple of a and b is a multiple of m.

Chapter I

SEMI-GROUPS AND GROUPS

The theory of groups is one of the oldest and richest branches of abstract algebra. Groups of transformations play an important role in geometry, and finite groups are fundamental in Galois' discoveries in the theory of equations. These two fields provided the original impetus to the development of the theory of groups.

A more general concept than that of a group is that of a semi-group. Though this notion appears to be useful in many connections, the theory of semi-groups is comparatively new and it certainly cannot be regarded as having reached a definitive stage. In this chapter we shall begin with this more general concept, but we treat it only briefly. Our aims in considering semi-groups are to provide an introduction to the theory of groups and to obtain some elementary results that will be useful in the study of rings. The main part of our discussion deals with groups. The principal concepts that we consider here are those of isomorphism, homomorphism, subgroup, invariant subgroup, factor group, and transformation group.

1. Definition and examples of semi-groups. We have defined a binary composition in a set $\mathfrak{S}$ to be a mapping of the product set $\mathfrak{S} \times \mathfrak{S}$ into the set $\mathfrak{S}$. The image in $\mathfrak{S}$ of the pair (a,b) in $\mathfrak{S} \times \mathfrak{S}$ is usually called the *product* or the *sum* of a and b. Accordingly, this result is denoted as $a \cdot b \equiv ab$ or as $a + b$. Occasionally other notations such as $a \circ b$, $a \times b$, $[a,b]$ are employed. In this book we shall be concerned almost exclusively with compositions that are *associative* in the sense that

$$(1) \qquad\qquad (ab)c = a(bc)$$

15

holds for all a,b,c in $\mathfrak{S}$. This concept is the essential ingredient in the algebraic system that we now define.

Definition 1. *A* semi-group *is a system* <u>*consisting*</u> *of a set* $\mathfrak{S}$ *and an associative binary composition in* $\mathfrak{S}$.

In describing a particular semi-group one has to specify the composition as well as the set $\mathfrak{S}$ in which it acts. Thus the same set may be the set part of many different semi-groups. Nevertheless for the sake of brevity we shall often call the set $\mathfrak{S}$ "the semi-group $\mathfrak{S}$." The precise terminology should, of course, be "the set $\mathfrak{S}$ of the semi-group," but in most instances there will be little likelihood of confusion in using the abbreviated phrase.

Examples. (1) The set P of positive integers and the composition of ordinary addition in P. (2) P and ordinary multiplication. (3) P and the composition $(a,b) \rightarrow a \cdot b \equiv a + b + ab$. It can be verified that this is associative. (4) The set I of integers, addition as composition. (5) I and multiplication. (6) The set $P(S)$ of subsets of a set, the join composition $(A,B) \rightarrow A \cup B$. (7) $P(S)$ and the intersection composition.

An important type of semi-group is obtained from the totality $\mathfrak{T}$ of transformations (single-valued mappings) of a given set S. We introduce in $\mathfrak{T}$ the mapping $(\alpha,\beta) \rightarrow \alpha\beta$ where, as usual, $\alpha\beta$ denotes the resultant of the transformations α and β. It is necessary to verify the associative law. More generally, we consider four sets S, T, U and V. Let α be a mapping of S into T, β a mapping of T into U and γ a mapping of U into V. The mappings $(\alpha\beta)\gamma$ and $\alpha(\beta\gamma)$ are defined. We now show that they are equal. Thus let x be any element of S. Then by definition $x((\alpha\beta)\gamma) = (x(\alpha\beta))\gamma = ((x\alpha)\beta)\gamma$ and $x(\alpha(\beta\gamma)) = (x\alpha)(\beta\gamma) = ((x\alpha)\beta)\gamma$. Hence $x((\alpha\beta)\gamma) = x(\alpha(\beta\gamma))$ for all x, and this is what is meant by saying that $(\alpha\beta)\gamma = \alpha(\beta\gamma)$. In particular we see that the associative law holds for the resultant of transformations of one set S.

As a special case of this type of semi-group let S be a finite set comprising n elements. We can take these to be the integers $1, 2, \cdots, n$. The mapping α may be denoted by the symbol

$$(2) \qquad \begin{pmatrix} 1 & 2 & 3 & \cdots & n \\ 1\alpha & 2\alpha & 3\alpha & \cdots & n\alpha \end{pmatrix}$$

in which the image $k\alpha$ of k is written below the element k. Clearly the number of mappings of S into itself is the number of distinct ways of writing the second line in (2). Since we have n choices for each of the places in the second line, the *order* or number of elements in $\mathfrak{T}$ is n^n.

A semi-group is said to be *finite* if it contains only a finite number of elements. In investigating such a semi-group it is useful to tabulate the products $\alpha\beta$ in a *multiplication table* for $\mathfrak{S}$. If $\alpha_1, \alpha_2, \cdots, \alpha_m$ are the elements of $\mathfrak{S}$ such a table has the form

	α_1	α_2	$\cdots$	α_j	$\cdots$	α_m
α_1				$\cdot$		
α_2				$\cdot$		
$\vdots$				$\vdots$		
α_i	$\cdot$	$\cdot$	$\cdots$	$\alpha_i\alpha_j$	$\cdots$	$\cdot$
$\vdots$				$\vdots$		
α_m				$\cdot$		

Here we write the product $\alpha_i\alpha_j$ in the intersection of the row containing α_i with the column containing α_j. For example let $\mathfrak{T}$ be the semi-group of transformations of a set of two elements. The elements of $\mathfrak{T}$ are

$$\epsilon = \begin{pmatrix} 1 & 2 \\ 1 & 2 \end{pmatrix}, \quad \alpha = \begin{pmatrix} 1 & 2 \\ 2 & 1 \end{pmatrix}, \quad \beta = \begin{pmatrix} 1 & 2 \\ 1 & 1 \end{pmatrix}, \quad \gamma = \begin{pmatrix} 1 & 2 \\ 2 & 2 \end{pmatrix}.$$

A multiplication table for $\mathfrak{T}$ is

	ϵ	α	β	γ
ϵ	ϵ	α	β	γ
α	α	ϵ	β	γ
β	β	γ	β	γ
γ	γ	β	β	γ

2. Non-associative binary compositions. We consider for a moment an arbitrary (not necessarily associative) binary composition $(a,b) \rightarrow ab$ in a set $\mathfrak{S}$. Such a mapping defines two *ternary compositions*, that is, mappings of $\mathfrak{S} \times \mathfrak{S} \times \mathfrak{S}$ into $\mathfrak{S}$. These are the mappings $(a,b,c) \rightarrow (ab)c$ and $(a,b,c) \rightarrow a(bc)$. More generally we can define inductively a number of n-ary compositions in $\mathfrak{S}$. Suppose that these have already been built up out of the binary composition to the stage of m-ary compositions for every $m < n$. It is understood here that for $m = 1$ the identity mapping $a \rightarrow a$ is taken. Now let m be any positive integer $< n$ and let

$$(a_1, a_2, \cdots, a_m) \rightarrow u(a_1, a_2, \cdots, a_m)$$

$$(a_{m+1}, a_{m+2}, \cdots, a_n) \rightarrow v(a_{m+1}, a_{m+2}, \cdots, a_n)$$

be definite m-ary and $(n - m)$-ary compositions determined by the original binary one. Then we take the mapping

$$(a_1, a_2, \cdots, a_n) \rightarrow u(a_1, a_2, \cdots, a_m)v(a_{m+1}, a_{m+2}, \cdots, a_n)$$

as one of our n-ary compositions. All the mappings obtained in this way by varying m, u and v are the n-ary compositions associated with $(a,b) \rightarrow ab$. The results of applying these mappings to $(a_1, a_2, \cdots, a_n)$ will be called (complex) *products* of $a_1, a_2, \cdots, a_n$ (taken in this order).

For example, the possible products of a_1, a_2, a_3, a_4 are

$$((a_1a_2)a_3)a_4, \ (a_1(a_2a_3))a_4, \ (a_1a_2)(a_3a_4), \ a_1(a_2(a_3a_4)), \ a_1((a_2a_3)a_4).$$

One can easily construct a set with a binary composition for which the indicated n-ary compositions are all distinct. For this purpose let S be a set with distinct elements $a_1, a_2, a_3, \cdots$ and let $\mathfrak{S}^*$ be the set of symbols that can be obtained as follows: Select any finite set of elements $a, b, \cdots, s$ in a definite order in the set S. If this set has either one or two elements then we include it in $\mathfrak{S}^*$. If it has more than two elements then we partition it into two ordered subsets $a, b, \cdots, k$ and $l, \cdots, s$ and we inclose the subsets thus obtained that contain more than one element in parentheses. This gives $(a, b, \cdots, k)(l, \cdots, s)$. We then repeat these rules on the two subsets and continue until the process terminates. If u and v represent any two symbols in $\mathfrak{S}^*$, then we define

$$uv = \begin{cases} uv \text{ if both } u \text{ and } v \text{ are in } S \\ u(v) \text{ if } u \; \varepsilon \; S \text{ and } v \text{ has more than one term} \\ (u)v \text{ if } v \; \varepsilon \; S \text{ and } u \text{ has more than one term} \\ (u)(v) \text{ if both } u \text{ and } v \text{ have more than one term.} \end{cases}$$

It is clear that this gives a binary composition in $\mathfrak{S}^*$. Moreover the n-ary compositions that we defined before are all different in $\mathfrak{S}^*$ since they give different results for the elements $a_1, a_2, \cdots, a_n$. If $N(n)$ denotes the number of these compositions, then our definition gives the recursion formula

$$(3) \quad N(n) = N(n-1)N(1) + N(n-2)N(2) + \cdots$$
$$+ N(1)N(n-1).$$

Also $N(1) = 1$. It is also clear that for any binary composition in any set, $N(n)$ is an upper bound for the number of distinct induced n-ary compositions.

It is easy to solve the recursion formula (3) and obtain an explicit formula for $N(n)$. For this purpose we introduce the "generating function" defined by the power series

$$y = N(1)x + N(2)x^2 + \cdots + N(n)x^n + \cdots.$$

Then

$$y^2 = N(1)N(1)x^2 + [N(2)N(1) + N(1)N(2)]x^3 + \cdots$$
$$= N(2)x^2 + N(3)x^3 + \cdots.$$

Since $N(1) = 1$, this gives

$$y^2 - y + x = 0.$$

Hence

$$y = \frac{1 - (1 - 4x)^{\frac{1}{2}}}{2} = \sum_1^\infty \frac{1 \cdot 3 \cdots (2n-3)}{1 \cdot 2 \cdots n} 2^{n-1} x^n$$

and

$$(4) \qquad N(n) = \frac{1 \cdot 3 \cdots (2n-3)}{1 \cdot 2 \cdots n} 2^{n-1}.{}^*$$

---

* This can be written more concisely as $N(n) = \dfrac{(2n-2)!}{n!(n-1)!}$.

EXERCISES

1. In the set I of integers define the binary composition $f(x,y) = x + y^2$. Work out all of the induced 4-ary compositions.

2. For a given binary composition define a *simple product* of n a's inductively as either $a_1 u$ where u is a simple product of $a_2, \cdots, a_n$ or $v a_n$ where v is a simple product of $a_1, \cdots, a_{n-1}$. Show that any product of $\geq 2^r$ elements can be regarded as a simple product of r elements (that are themselves products).

3. Generalized associative law. Powers.

We shall now show that if our binary composition is associative then all the possible products of $a_1, a_2, \cdots, a_n$ taken in this order are equal. We first define a particular product $\prod_1^m a_i$ by the formulas

$$\prod_1^1 a_i = a_1, \qquad \prod_1^{r+1} a_i = \left(\prod_1^r a_j \right) a_{r+1}$$

and we prove the

Lemma. $\displaystyle \prod_1^n a_i \prod_1^m a_{n+j} = \prod_1^{n+m} a_k.$

Proof. By definition this holds if $m = 1$. Assume it true for $m = r$ and consider the case $m = r + 1$. Here

$$\prod_1^n a_i \prod_1^{r+1} a_{n+j} = \prod_1^n a_i \left(\left(\prod_1^r a_{n+j} \right) a_{n+r+1} \right)$$

$$= \left(\prod_1^n a_i \prod_1^r a_{n+j} \right) a_{n+r+1}$$

$$= \left(\prod_1^{n+r} a_k \right) a_{n+r+1}$$

$$= \prod_1^{n+r+1} a_k.$$

Consider now any product associated with $(a_1, a_2, \cdots, a_n)$. By definition it is a product uv where u is a product associated with $(a_1, a_2, \cdots, a_m)$, $1 < m < n$, and v is a product associated with $(a_{m+1}, \cdots, a_n)$. By induction we can assume that $u = \prod_{i=1}^m a_i$ and $v = \prod_{j=1}^{n-m} a_{m+j}$. Hence $uv = \prod_{k=1}^n a_k$. Thus all products determined by $(a_1, a_2, \cdots, a_n)$ are equal. From now on we shall denote this uniquely determined product as $a_1 a_2 \cdots a_n$ omitting all parentheses.

If all the $a_i = a$, we denote $a_1 a_2 \cdots a_n$ by a^n and call this element the nth *power* of a. Our remarks show that

$$(5) \qquad a^n a^m = a^{n+m}, \quad (a^n)^m = a^{nm}.$$

If the notation $+$ is used for the composition in $\mathfrak{S}$, then we write

$$a_1 + a_2 + \cdots + a_n \text{ in place of } a_1 a_2 \cdots a_n,$$

$$na \text{ in place of } a^n.$$

The rules (5) for powers now become the following rules for *multiples na*:

$$(5') \qquad na + ma = (n + m)a, \quad m(na) = (mn)a.$$

4. Commutativity. If a and b are elements of a semi-group it may happen that $ab \neq ba$. For example, in the semi-group whose multiplication table is given in §1 we have $\alpha\beta = \beta$ whereas $\beta\alpha = \gamma$. If $ab = ba$ in $\mathfrak{S}$, then the elements a and b are said to *commute* and if this holds for any pair a,b in $\mathfrak{S}$ then $\mathfrak{S}$ is called *commutative*. It is immediate by induction on n that if $a_i b = ba_i$, $i = 1, 2, \cdots, n$, then

$$a_1 \cdots a_n b = ba_1 \cdots a_n.$$

Suppose next that for the elements $a_1, a_2, \cdots, a_n$ we have the commutativity $a_i a_j = a_j a_i$ for all i, j and consider any product $a_{1'} a_{2'} \cdots a_{n'}$ where $1', 2', \cdots, n'$ is some permutation of the numbers $1, 2, \cdots, n$. Suppose that a_n occurs in the hth place in this product. Then $a_{h'} = a_n$. Hence

$$a_{1'} a_{2'} \cdots a_{h'} \cdots a_{n'} = a_{1'} \cdots a_{(h-1)'} a_{(h+1)'} \cdots a_{n'} a_n.$$

Using induction, we may assume that

$$a_{1'} \cdots a_{(h-1)'} a_{(h+1)'} \cdots a_{n'} = a_1 a_2 \cdots a_{n-1}.$$

Hence $a_{1'} a_{2'} \cdots a_{n'} = a_1 a_2 \cdots a_n$.

The powers of a single element commute since (5) holds. Also it is clear from our discussion that if $ab = ba$, then

$$(6) \qquad (ab)^n = a^n b^n.$$

In the additive notation this reads

$$(6') \qquad\qquad n(a + b) = na + nb.$$

5. Identities and inverses. An element e of a semi-group $\mathfrak{S}$ is called a *left identity* (*unit, unity*) if $ea = a$ for every a in $\mathfrak{S}$. Similarly f is a *right identity* if $af = a$ for every a.

Examples. (1) The semi-group of positive integers relative to multiplication has the two-sided (= left and right) identity 1. (2) The semi-group of positive integers relative to addition has no identity. (3) Let $\mathfrak{S}$ be any set and define in $\mathfrak{S}$, $ab = b$. Then $\mathfrak{S}$ is a semi-group and any element of $\mathfrak{S}$ is a left identity. On the other hand, if $\mathfrak{S}$ possesses more than one element, then it has no right identities.

The last example shows that a semi-group can have several left (right) identities but no right (left) identities. However, if $\mathfrak{S}$ possesses a left identity e and a right identity f, then necessarily $e = f$; for $ef = f$ since e is a left identity and $ef = e$ since f is a right identity. This shows that, if we have a left identity and a right identity, then we cannot have more than one of either type. In particular, if a two-sided identity exists, then it is unique.

From now on we refer to a two-sided identity simply as an (the) *identity* and we shall usually denote this element as 1. An element a of $\mathfrak{S}$ will be called *right regular* if there exists an a' in $\mathfrak{S}$ such that $aa' = 1$. The element a' is called a *right inverse* of a. Left regularity and left inverses are defined in a similar manner. If a is both left regular and right regular, then we shall say that it is a *unit* (*regular*). In this case we have an a' such that $aa' = 1$ and an a'' such that $a''a = 1$. Then

$$a' = (a''a)a' = a''(aa') = a''.$$

Thus $a' = a''$ and this element is called an *inverse* of a. Our argument shows that it is unique. We shall denote this element as a^{-1}. Since $aa^{-1} = 1 = a^{-1}a$, it is clear that a^{-1} is regular and that a is its inverse. This is the rule: $(a^{-1})^{-1} = a$. We note also that, if a and b are units, then so is ab since $(ab)(b^{-1}a^{-1}) = 1 = (b^{-1}a^{-1})(ab)$. Thus we have $(ab)^{-1} = b^{-1}a^{-1}$.

If the operation in $\mathfrak{S}$ is denoted as $+$, we denote the identity as 0. The inverse of a if it exists is written as $-a$. Thus we have $-(-a) = a$ and $-(a + b) = -b + (-a)$. Also we shall write $a - b$ for $a + (-b)$.

6. Definition and examples of groups.

Definition 2. *A group is a semi-group that has an identity and in which every element is a unit.*

Thus a group is a system consisting of a set $\mathfrak{G}$ and binary composition in $\mathfrak{G}$ such that the following conditions hold:

1. $$(ab)c = a(bc).$$

2. There exists an element 1 in $\mathfrak{G}$ such that $a1 = a = 1a$.

3. For each a in $\mathfrak{G}$ there is an element a^{-1} in $\mathfrak{G}$ such that $aa^{-1} = 1 = a^{-1}a$.

As in the case of semi-groups we shall often use the term "group $\mathfrak{G}$" for the set part of the group. The following is a list of examples of groups all of which should be familiar to the reader.

Examples. (1) R_+, the totality of real numbers, addition as composition. Here the number 0 is the identity and the inverse of a is the usual $-a$. (2) C_+, the set of complex numbers, addition as composition. (3) R^*, the set of non-zero real numbers, multiplication as the composition. Here the real number 1 is the identity and the inverse of a is the usual reciprocal a^{-1}. (4) Q, the set of positive real numbers, ordinary multiplication. (5) C^*, the set of non-zero complex numbers, multiplication. (6) U, the set of complex numbers $e^{i\theta}$ of absolute value 1, multiplication. (7) U_n, the n complex nth roots of 1, multiplication. (8) The totality of rotations about a point O in the plane, composition the resultant. If O is taken to be the origin, the rotation through an angle θ can be represented analytically as the mapping $(x,y) \to (x',y')$ where

$$x' = x \cos \theta - y \sin \theta, \quad y' = x \sin \theta + y \cos \theta.$$

If $\theta = 0$, we get the identity transformation and this acts as the identity in the set of rotations. The inverse of the rotation through the angle θ is the rotation through the angle $-\theta$. (9) The totality of rotations about a point O in space, resultant composition. (10) The set of vectors in the plane, vector addition as composition. Analytically a vector may be represented as a pair of real numbers (a,b). These are respectively the x- and the y-coordinates of the vector. If $v = (a,b)$ and $v' = (a',b')$, the usual vector addition gives $v + v' = (a + a', b + b')$. The 0 vector $0 = (0,0)$ acts as the identity and the inverse of v is $-v = (-a,-b)$.

EXERCISE

1. Let $\mathfrak{G}$ be the totality of pairs of real numbers (a,b) for which $a \neq 0$. Take the composition in $\mathfrak{G}$ that is defined by the formula

$$(a,b)(c,d) = (ac, bc + d).$$

Verify that this is a group.

It is clear from our discussion of semi-groups that the identity element is unique in $\mathfrak{G}$. Also the inverse of a is uniquely determined. If a and b are any two elements of a group $\mathfrak{G}$ then the linear equation $ax = b$ has the solution $a^{-1}b$ in $\mathfrak{G}$. This is the only solution since $ax = ax'$ implies that $a^{-1}(ax) = a^{-1}(ax')$. Hence $x = x'$. This last remark shows that the *left cancellation law* holds. Similarly the equation $ya = b$ has a unique solution in $\mathfrak{G}$ and the right cancellation law holds. The solvability of $ax = b$ and $ya = b$ in $\mathfrak{G}$ is a characteristic property of a group (see ex. 3 below).

EXERCISES

1. An element e of a semi-group is said to be *idempotent* if $e^2 = e$. Show that the only idempotent element in a group is $e = 1$.

2. Prove that a semi-group having the following properties is a group:

(a) $\mathfrak{G}$ has a right identity 1_r.

(b) Every element a of $\mathfrak{G}$ has a right inverse relative to 1_r.*

3. Prove that if $\mathfrak{G}$ is a semi-group in which the equations $ax = b$ and $ya = b$ are solvable for any a and b, then $\mathfrak{G}$ is a group.

4. Prove that a finite semi-group in which the cancellation laws hold is a group.

7. Subgroups. A subset $\mathfrak{S}'$ of a semi-group is said to be *closed* if $ab \, \varepsilon \, \mathfrak{S}'$ for every a and b in $\mathfrak{S}'$. It is clear that the associative law holds in $\mathfrak{S}'$. Hence the pair $\mathfrak{S}', \cdot$ consisting of $\mathfrak{S}'$ and the induced mapping $(a,b) \rightarrow ab$, a,b in $\mathfrak{S}'$, form a semi-group. We call such a semi-group a *sub-semi-group* of the given semi-group. It may happen that $\mathfrak{S}'$ is a group relative to the composition in $\mathfrak{S}$. In this case we say that $\mathfrak{S}'$ is a *subgroup* of $\mathfrak{S}$.

Examples. (1) The set of positive integers is (strictly speaking, *determines*) a sub-semi-group of the group I_+ of integers relative to addition. The set of even integers is a subgroup of I_+. More generally the totality of multiples km of a fixed integer m is a subgroup. (2) The set consisting of the numbers 1 and -1 is a subgroup of the semi-group of integers relative to multiplication.

We shall show now that, if $\mathfrak{S}$ is any semi-group with an identity, then the subset $\mathfrak{G}$ of units of $\mathfrak{S}$ determines a subgroup. Let a and b be units; then we have seen that $b^{-1}a^{-1}$ is an inverse for ab. Hence $ab \, \varepsilon \, \mathfrak{G}$. Since $1 \cdot 1 = 1$, $1 \, \varepsilon \, \mathfrak{G}$ and this element acts as an

---

identity in $\mathfrak{G}$. Finally, if $a \,\varepsilon\, \mathfrak{G}$, then $a^{-1} \,\varepsilon\, \mathfrak{G}$ since $aa^{-1} = 1 = a^{-1}a$. Thus every element of $\mathfrak{G}$ has an inverse in $\mathfrak{G}$. We shall call $\mathfrak{G}$ *the group of units* of $\mathfrak{S}$. The example (2) given above is the group of units in the semi-group of integers under multiplication. We shall see in the sequel that many important examples of groups are obtained as groups of units of semi-groups.

We begin next with an arbitrary group $\mathfrak{G}$ and we shall determine the conditions that a subset $\mathfrak{H}$ of $\mathfrak{G}$ determines a subgroup of $\mathfrak{G}$. First we know that $\mathfrak{H}$ must be closed. Next $\mathfrak{H}$ has an identity $1'$. Since $(1')^2 = 1'$, it is clear (ex. 1, p. 24) that $1' = 1$, the identity of $\mathfrak{G}$. Finally, if $a \,\varepsilon\, \mathfrak{H}$, then there exists an element a' in $\mathfrak{H}$ such that $aa' = 1 = a'a$. Then a' is an inverse of a and since there is only one inverse, $a' = a^{-1}$. This shows that the following conditions are necessary in order that a subset $\mathfrak{H}$ of a group $\mathfrak{G}$ determines a subgroup of $\mathfrak{G}$:

1. $a,b \,\varepsilon\, \mathfrak{H}$ implies that $ab \,\varepsilon\, \mathfrak{H}$ (closure).
2. $1 \,\varepsilon\, \mathfrak{H}$.
3. $a \,\varepsilon\, \mathfrak{H}$ implies that $a^{-1} \,\varepsilon\, \mathfrak{H}$.

These conditions are also sufficient conditions on a subset $\mathfrak{H}$ that $\mathfrak{H}, \cdot$ be a subgroup of $\mathfrak{G}, \cdot$; for it is clear that they imply axioms 2 and 3 for a group. Moreover, the associativity condition certainly holds in $\mathfrak{H}$ since it holds in $\mathfrak{G}$.

It should be noted that the group $\mathfrak{G}$ itself can be regarded as a subgroup of $\mathfrak{G}$. If $\mathfrak{H}$ is a subgroup and $\mathfrak{H}$ is a proper subset of $\mathfrak{G}$, then we say that $\mathfrak{H}$ is a *proper subgroup of* $\mathfrak{G}$. We remark also that the subset of $\mathfrak{G}$ consisting of the element 1 only is a subgroup. This is evident from the definition or from the foregoing conditions. We shall denote this subgroup as the subgroup 1 of $\mathfrak{G}$ (or 0 in the additive notation).

EXERCISES

1. Verify that the subset of pairs of the form $(1,b)$ forms a subgroup of the group given in ex. 1, p. 23.

2. Show that a non-vacuous subset $\mathfrak{H}$ of a group $\mathfrak{G}$ is a subgroup if and only if $ab^{-1} \,\varepsilon\, \mathfrak{H}$ for any a and b in $\mathfrak{H}$.

3. Prove that any finite sub-semi-group of a group is a subgroup (cf. ex. 4, p. 24).

4. Prove that, if A is any collection of subgroups $\mathfrak{H}$ of $\mathfrak{G}$, then the intersection $\bigcap_A \mathfrak{H}$ is a subgroup.

5. Prove that, if a is any element of a group $\mathfrak{G}$, then the set $\mathfrak{C}(a)$ of elements that commute with a is a subgroup of $\mathfrak{G}$.

8. Isomorphism. We shall consider first a well-known example of this fundamental concept. Let R_+ be the group of real numbers relative to addition and let Q be the group of positive real numbers relative to multiplication. We consider the mapping $x \rightarrow e^x$ of R_+ into Q. This mapping is 1–1 of R_+ onto Q and its inverse is the mapping $z \rightarrow \log z$. Also we have the fundamental property:

$$e^{x+y} = e^x e^y.$$

Thus we arrive at the same result if (a) we first perform the group composition on two numbers in R_+ and then take the image in Q, or (b) we first take images in Q and then perform the group composition on these images. From the abstract point of view the groups R_+ and Q are essentially indistinguishable; for we are not interested in the nature of the elements of our groups but only in their compositions and these are essentially the same in the two examples. The precise relation between R_+ and Q can be stated by saying that these two groups are isomorphic in the sense of the following

Definition 3. *Two groups $\mathfrak{G}$ and $\mathfrak{G}'$ are said to be* isomorphic *if there exists a 1–1 mapping $x \rightarrow x'$ of $\mathfrak{G}$ onto $\mathfrak{G}'$ such that $(xy)' = x'y'$.*

A mapping satisfying the condition of this definition is called an *isomorphism* of $\mathfrak{G}$ onto $\mathfrak{G}'$. If $\mathfrak{G}$ and $\mathfrak{G}'$ are isomorphic, there may exist many isomorphisms between them. For example, if a is any positive number $\neq 1$, then the mapping $x \rightarrow a^x$ is an isomorphism of R_+ onto Q. Isomorphic groups are often said to be *abstractly equivalent.* If $\mathfrak{G}$ is isomorphic to $\mathfrak{G}'$, we write $\mathfrak{G} \cong \mathfrak{G}'$. It is clear that the isomorphism relation between groups is an equivalence; for the identity mapping is an isomorphism of $\mathfrak{G}$ onto itself and, if $a \rightarrow a'$ is an isomorphism of $\mathfrak{G}$ onto $\mathfrak{G}'$, then $a' \rightarrow a$, the inverse mapping, is an isomorphism of $\mathfrak{G}'$ onto $\mathfrak{G}$. Finally, if $a \rightarrow a'$ is an isomorphism of $\mathfrak{G}$ onto $\mathfrak{G}'$

and $a' \to a''$ is an isomorphism of $\mathfrak{G}'$ onto $\mathfrak{G}''$, then $a \to a''$ is an isomorphism of $\mathfrak{G}$ onto $\mathfrak{G}''$.

EXERCISES

1. Prove that, if $x \to x'$ is an isomorphism, then $1'$, the image of 1, is the identity of the second group. Prove also that $(a^{-1})' = (a')^{-1}$.

2. Is the mapping $\theta \to e^{i\theta}$ an isomorphism of R_+ onto the multiplicative group of complex numbers of absolute value 1?

9. Transformation groups. Let S be an arbitrary set and let $\mathfrak{T}(S)$ be the semi-group of transformations of S into itself. We know that $\mathfrak{T}$ has an identity, namely, the identity mapping $x \to x$. We consider now the subgroup $\mathfrak{G}(S)$ of units of $\mathfrak{T}(S)$. We shall show that $\mathfrak{G}(S)$ is just the set of 1–1 mappings of S onto itself; for we have seen that, if α is 1–1 of S onto S, then the inverse mapping α^{-1} has the property $\alpha\alpha^{-1} = 1 = \alpha^{-1}\alpha$. On the other hand, let α be any element of $\mathfrak{T}(S)$ for which there exists an inverse β such that $\alpha\beta = 1 = \beta\alpha$. Then any $x = (x\beta)\alpha \ \varepsilon \ S\alpha$ so that α maps S onto itself. Also, if $x\alpha = y\alpha$, then $(x\alpha)\beta = (y\alpha)\beta$ and $x = y$. Hence α is 1–1. We shall call $\mathfrak{G}(S)$ *the group of* 1–1 *transformations* or *permutations* of the set S.

More generally, we define *a transformation group* (in S) to be any subgroup of a group $\mathfrak{G}(S)$. If we recall the conditions that a subset $\mathfrak{H}$ be a subgroup, we see that a set $\mathfrak{H}$ of 1–1 transformations of a set S onto itself determines a transformation group if the following hold:

1. If $\alpha, \beta \ \varepsilon \ \mathfrak{H}$, then the resultant $\alpha\beta \ \varepsilon \ \mathfrak{H}$.
2. The identity mapping $x \to x$ is in $\mathfrak{H}$.
3. If $\alpha \ \varepsilon \ \mathfrak{H}$, the inverse mapping α^{-1} is in $\mathfrak{H}$.

We consider now the special case in which S is the set of n numbers 1, 2, $\cdots$, n. The group $\mathfrak{G}(S)$ of permutations of S is called the *symmetric group of degree* n. It is usually denoted as S_n. We shall represent an element $\alpha \ \varepsilon \ S_n$ by a symbol of the form

$$\begin{pmatrix} 1 & 2 & \cdots & n \\ 1\alpha & 2\alpha & \cdots & n\alpha \end{pmatrix}$$

and we can use this representation to calculate the order (number of elements) of the group S_n. Clearly the element 1α is arbitrary.

Hence we can choose the number in the first position in n different ways. Since no repetitions are allowed in the second row of our symbol, we have $n - 1$ choices for the second position, $n - 2$ for the third, etc. Hence in all we have $n!$ symbols and consequently $n!$ elements in S_n.

EXERCISES

1. Calculate $\alpha\beta$, $\beta\alpha$ and α^{-1} if

$$\alpha = \begin{pmatrix} 1 & 2 & 3 & 4 & 5 \\ 2 & 3 & 1 & 5 & 4 \end{pmatrix}, \quad \beta = \begin{pmatrix} 1 & 2 & 3 & 4 & 5 \\ 1 & 3 & 4 & 5 & 2 \end{pmatrix}.$$

2. Write down the elements of S_3 and work out a multiplication table for this group.

3. Verify that the transformations

$$\begin{pmatrix} 1 & 2 & 3 \\ 1 & 2 & 3 \end{pmatrix}, \quad \begin{pmatrix} 1 & 2 & 3 \\ 2 & 3 & 1 \end{pmatrix}, \quad \begin{pmatrix} 1 & 2 & 3 \\ 3 & 1 & 2 \end{pmatrix}$$

form a transformation group.

4. Which of the examples given in § 6 are transformation groups?

5. Verify that the set of transformations of the line given by the rule $x \rightarrow ax + b$, $a \neq 0$ form a transformation group. Show that this group is isomorphic to the one given in ex. 1, p. 23.

6. Verify that the totality of transformations of the plane defined by $(x, y) \rightarrow (x + a, 0)$ constitute a group relative to resultant composition. Is this a transformation group?

10. Realization of a group as a transformation group. Historically the theory of groups dealt at first only with transformation groups. The concept of an abstract group was introduced later for the purpose of deriving in the simplest and most direct manner those properties of transformation groups that concern the resultant composition only and do not refer to the set S in which the transformations act. It is natural to ask whether or not the abstract concept is completely appropriate in the sense that the class of systems covered by it is just the class of transformation groups. This question is answered affirmatively in the following fundamental theorem due to Cayley:

Theorem 1. *Any group is isomorphic to a transformation group.*

Proof. The transformation group that we shall define will act in the set $\mathfrak{G}$ of the given group. With each element a of the group $\mathfrak{G}$ we associate the mapping

$$x \to xa$$

of the set $\mathfrak{G}$ into itself. We denote this mapping as a_r and call it the *right multiplication* determined by a. Since the right cancellation law holds, a_r is 1-1. Since any b can be written in the form $(ba^{-1})a = (ba^{-1})a_r$, a_r is a mapping onto $\mathfrak{G}$. Hence a_r is in the group of 1-1 transformations of the set $\mathfrak{G}$. We wish to show now that the totality $\mathfrak{G}_r = \{a_r\}$ is a transformation group in $\mathfrak{G}$. Consider first the product $a_r b_r$. This sends x into $(xa)b$. By the associative law $(xa)b = x(ab)$. Thus $a_r b_r$ has the same effect as $(ab)_r$. Hence

$$(7) \qquad\qquad a_r b_r = (ab)_r$$

is in $\mathfrak{G}_r$. We note next that $1 = 1_r$ is in $\mathfrak{G}_r$. Finally by (7) $a_r(a^{-1})_r = 1_r = (a^{-1})_r a_r$. Hence $a_r^{-1} = (a^{-1})_r$ is in $\mathfrak{G}_r$. Thus $\mathfrak{G}_r$ is a transformation group. We consider now the correspondence $a \to a_r$ of the group $\mathfrak{G}$ onto the group $\mathfrak{G}_r$. If $a \neq b$, then $1a_r = a \neq b = 1b_r$. Hence $a_r \neq b_r$. Thus $a \to a_r$ is 1-1. Since (7) holds, the mapping $a \to a_r$ is an isomorphism. This completes the proof.

We shall refer to the isomorphism $a \to a_r$ as the (right) *regular realization* of $\mathfrak{G}$ as a transformation group. It should be observed that if $\mathfrak{G}$ is a finite group of order n, then $\mathfrak{G}_r$ is a subgroup of the symmetric group S_n. Hence we have the

Corollary. *Any finite group of order n is isomorphic to a subgroup of S_n.*

Examples. (1) R_+, the group of real numbers and addition. If $a \in R_+$, a_r is the translation $x \to x' = x + a$. (2) R^*, the group of real numbers $\neq 0$ under multiplication. Here a_r is the dilation $x \to x' = ax$. (3) The group of pairs of real numbers (a,b), $a \neq 0$, where $(a,b)(c,d) = (ac,\ bc + d)$. Here $(c,d)_r$ maps (x,y) into (x',y') where

$$x' = cx, \quad y' = cy + d.$$

There is a second realization of $\mathfrak{G}$ as a transformation group that one obtains by using left multiplications. We define the *left multiplication* a_l as the mapping $x \to ax$ of $\mathfrak{G}$ into itself. As in the case of right multiplication it is easy to see that a_l is 1-1 of $\mathfrak{G}$ onto itself. Also the set $\mathfrak{G}_l$ of the a_l is a transformation group. The proof of this is the same as for $\mathfrak{G}_r$ with the modification that

$$(8) \qquad\qquad a_l b_l = (ba)_l.$$

This follows from

$$x a_l b_l = b(ax) = (ba)x = x(ba)_l.$$

The mapping $a \to a_l$ is 1–1 of $\mathfrak{G}$ onto $\mathfrak{G}_l$ but in general this is not an isomorphism. In order to obtain an isomorphism we must replace this mapping by the mapping $a \to a_l^{-1} = (a^{-1})_l$; for then we have

$$(ab)_l^{-1} = (b_l a_l)^{-1} = a_l^{-1} b_l^{-1}.$$

We shall call the isomorphism $a \to a_l^{-1}$ the *left regular realization* of $\mathfrak{G}$.

The associative law in $\mathfrak{G}$ gives the rule $a_l b_r = b_r a_l$ for all a,b in $\mathfrak{G}$ since $x a_l b_r = (ax)b$ and $x b_r a_l = a(xb)$. Hence any transformation belonging to the set $\mathfrak{G}_r$ commutes with any transformation belonging to $\mathfrak{G}_l$. The converse holds also, namely, if β is any transformation in $\mathfrak{G}$ that commutes with all the a_l (a_r), then β is a right (left) multiplication; for we have

$$x\beta = (x1)\beta = (1x_l)\beta = (1\beta)x_l = x(1\beta) = xb$$

for $b = 1\beta$. Hence $\beta = b_r$.

EXERCISE

1. Obtain the regular realizations of S_3.

11. Cyclic groups. Order of an element. Let M be any non-vacuous subset of a group $\mathfrak{G}$ and let $\{\mathfrak{H}\}$ be the collection of subgroups of $\mathfrak{G}$ that contain the set M. The collection $\{\mathfrak{H}\}$ contains $\mathfrak{G}$; hence it is not vacuous. Its intersection $\cap \mathfrak{H}$ is a subgroup of $\mathfrak{G}$ (ex. 4, p. 26). We denote this subgroup as $[M]$ and shall call it the *subgroup of $\mathfrak{G}$ generated by the set M*. The set $[M]$ has the following properties: (1) $[M]$ is a subgroup of $\mathfrak{G}$. (2) $[M] \supseteq M$. (3) If $\mathfrak{H}$ is any subgroup of $\mathfrak{G}$ containing M, then $\mathfrak{H} \supseteq [M]$. Also it is clear that these properties characterize $[M]$. Thus let $\mathfrak{K}$ be a subset of $\mathfrak{G}$ satisfying (1), (2) and (3) (for M).

Then since $\mathfrak{K}$ is a subgroup containing M, $\mathfrak{K} \supseteq [M]$. By symmetry $[M] \supseteq \mathfrak{K}$. Hence $\mathfrak{K} = [M]$.

We can use this characterization to obtain explicitly the elements of $[M]$. We assert that these are just the finite products $a_1 a_2 \cdots a_n$ (n arbitrary) where $a_i \, \varepsilon \, M$ or a_i is the inverse of an element of M. Let $\mathfrak{K}$ denote the collection of these products. Then it is immediate that $\mathfrak{K}$ is a subgroup of $\mathfrak{G}$ containing M. On the other hand, if $\mathfrak{H}$ is a subgroup of $\mathfrak{G}$ containing M, $\mathfrak{H}$ contains every $a \, \varepsilon \, M$ and every a^{-1} with a in M. Hence $\mathfrak{H}$ contains $\mathfrak{K}$. Thus $\mathfrak{K}$ satisfies (1), (2) and (3) and therefore $\mathfrak{K} = [M]$.

We consider now the special case in which $M = \{a\}$ is a set consisting of a single element a. Here we write $[a]$ for $[M]$, and we call this subgroup the *(cyclic) group generated by* a. A group $\mathfrak{Z}$ is called a *cyclic group* if there exists an $a \, \varepsilon \, \mathfrak{Z}$ such that $\mathfrak{Z} = [a]$. The element a is then called a *generator* of $\mathfrak{Z}$. The remark above shows that $[a]$ consists of the elements a^n, $n > 0$, 1 and $(a^{-1})^n$, $n > 0$. We shall now define $a^0 = 1$ and $a^{-n} = (a^{-1})^n$ if $n > 0$. In this sense $[a]$ consists of the integral powers of the element a.

A consideration of cases can be used to extend the basic laws of exponents (5) to all integral powers. For example, suppose $n > |m|$ and $m < 0$. Then $a^n a^m = a^n a^{-|m|} = a^n (a^{-1})^{|m|} = a^{n-|m|} = a^{n+m}$. We leave it to the reader to verify the other cases. We remark that by the laws of exponents, or directly, $[a]$ is a commutative group. The following are some familiar examples of cyclic groups.

Examples. (1) Let I_+ be the group of integers relative to addition. It is clear by the axiom of induction that a set of positive integers that contains 1 and that is closed under addition contains all the positive integers. From this it follows that $I_+ = [1]$. It is clear also that $I_+ = [-1]$ and that $1 \notin [k]$ if $k \neq 1, -1$. Hence 1 and -1 are the only generators of I.

(2) Let U_n be the group of complex nth roots of 1. Then U_n consists of the complex numbers $e^{\frac{2k\pi}{n}i}$, $k = 0, 1, 2, \cdots, n - 1$. Using the standard geometric representation of complex numbers, we see that these numbers are represented as the vertices of the regular n-gon inscribed in the unit circle that has (1,0) as one of its vertices. If we set $e^{\frac{2\pi}{n}i} = \rho$, we see that the elements of U_n are $1, \rho, \rho^2, \cdots, \rho^{n-1}$. Hence U_n is a cyclic group of order n.

Let $\mathfrak{Z}$ be a cyclic group with generator a and consider the mapping $n \to a^n$ of I_+ onto $\mathfrak{Z}$. This correspondence has the property

$$m + n \rightarrow a^{m+n} = a^m a^n.$$

Hence, if our mapping is 1–1, then it is an isomorphism of I_+ onto $\mathfrak{Z}$.

Suppose next that the mapping is not 1–1. Then $a^m = a^n$ for $m \neq n$. We may assume $n > m$. Then $a^{n-m} = a^n a^{-m} = a^m a^{-m} = 1$. Hence there exist positive integers p such that $a^p = 1$. Let r be the smallest positive integer having this property. Then we assert that the elements $1, a, \cdots, a^{r-1}$ are distinct and that every element of $\mathfrak{Z}$ is in this set; for if $a^k = a^l$ for $k \neq l$ and k, l in the range $0, 1, \cdots, r-1$, then $a^p = 1$ for $0 < p < r$ contrary to the choice of r. Next let a^n be any element of $\mathfrak{Z}$. Write $n = qr + s$, $0 \leq s < r$. Then $a^n = a^{qr+s} = a^{qr}a^s = (a^r)^q a^s = a^s$. This proves our assertion. Thus $\mathfrak{Z}$ is a finite group of order r.

We now see that if $\mathfrak{Z}$ is infinite the mapping $n \rightarrow a^n$ is necessarily 1–1. Hence any infinite cyclic group is isomorphic to I_+ and consequently any two infinite cyclic groups are isomorphic. We shall show next that any two cyclic groups of the same finite order are isomorphic. Let $\mathfrak{Z} = [a]$ and $\mathfrak{W} = [b]$ be of order r. We have seen that the order r of $[a]$ (or of $[b]$) is the smallest positive integer such that $a^r = 1$ ($b^r = 1$). We shall now show that, if h is any integer such that $a^h = 1$, then $r \mid h$. Thus suppose $h = rq + s$, $0 \leq s < r$. Then $a^h = 1$ gives $a^s = a^s 1^q = a^s (a^r)^q = a^{s+rq} = a^h = 1$. Hence $s = 0$ by the minimality of r. Now suppose that $a^n = a^m$. Then $a^{n-m} = 1$ and so $n - m = rq$. Hence $1 = b^{rq} = b^{n-m}$ and $b^n = b^m$. We can now map $a^n \rightarrow b^n$ and be sure that this correspondence is single-valued. By symmetry $b^n = b^m$ implies that $a^n = a^m$. Hence our mapping is 1–1. Clearly $a^n a^m = a^{n+m} \rightarrow b^{n+m} = b^n b^m$. Hence $a^n \rightarrow b^n$ is an isomorphism. This completes the proof of the following

Theorem 2. *Any two cyclic groups of the same order are isomorphic.*

The concept of a cyclic group gives us a first classification of the elements of an arbitrary group $\mathfrak{G}$. If a is any element of $\mathfrak{G}$, then we say that a is of *infinite order* or of *finite order* r, according as $[a]$ is infinite or is a finite group of order r. In the first case we know that $a^n \neq 1$ if n is any integer $\neq 0$, and if the second alternative holds, then $a^r = 1$. Also we know that r is the least positive integer such that $a^r = 1$.

Cyclic groups are the simplest kinds of groups. It is therefore not surprising that most questions concerning groups are readily answered for this type. Thus, for example, it is generally a very difficult task to determine all the subgroups of a given group. We shall now see that this can be done very simply for cyclic groups.

Let $\mathfrak{W}$ be a subgroup of the cyclic group $\mathfrak{Z} = [a]$. Assume first that $\mathfrak{W} \neq 1$. Then there exist positive integers m such that $a^m \varepsilon \mathfrak{W}$; for there exist integers $m \neq 0$ such that $a^m \varepsilon \mathfrak{W}$, and if $a^m \varepsilon \mathfrak{W}$, then so does $(a^m)^{-1} = a^{-m}$. Now let s be the smallest positive integer such that $a^s \varepsilon \mathfrak{W}$. We propose to show that $\mathfrak{W} = [a^s]$ and that the correspondence $\mathfrak{W} \to s$ is 1–1. To prove these results let $c = a^m$ be any element in $\mathfrak{W}$ and write $m = sq + u$ where $0 \leq u < s$. Then $a^u = a^m(a^s)^{-q} \varepsilon \mathfrak{W}$. Hence, by the minimality of s, $u = 0$. Thus $c = a^m = (a^s)^q$ and $\mathfrak{W} = [a^s]$. Also the 1–1 ness is clear since, if $\mathfrak{W} \to s$ and $\mathfrak{W}' \to s$, then $\mathfrak{W} = [a^s] = \mathfrak{W}'$.

If $\mathfrak{Z}$ is an infinite cyclic group, then our mapping $\mathfrak{W} \to s$ is a mapping onto the set of positive integers; for if we take any positive integer s, then $[a^s] \to s$ since the smallest positive integer p such that $a^p \varepsilon [a^s]$ is s itself.

Suppose next that $\mathfrak{Z}$ is finite of order r. Then we shall show that the mapping $\mathfrak{W} \to s$ is a mapping onto the set of positive integers $< r$ which are divisors of r. Since $1 = a^r \varepsilon \mathfrak{W}$, the argument used before shows that r is a multiple of s, that is, $s \mid r$. On the other hand, let s be any divisor of r and write $r = st$. Then $(a^s)^t = 1$, but $(a^s)^{t'} \neq 1$ if $0 < t' < t$. Hence, t is the order of $[a^s]$. Now if s' is the smallest positive integer such that $a^{s'} \varepsilon [a^s]$, then also $r = s't$ since $[a^{s'}] = [a^s]$. It follows that $s = s'$. Hence $[a^s] \to s$.

We have therefore proved the following

Theorem 3. *Let $\mathfrak{Z}$ be a cyclic group with generator a and let $\mathfrak{W}$ be any subgroup $\neq 1$ of $\mathfrak{Z}$. Then if s is the smallest positive integer such that $a^s \varepsilon \mathfrak{W}$, $\mathfrak{W} = [a^s]$. If $\mathfrak{Z}$ is infinite, then the correspondence $\mathfrak{W} \to s$ is a 1–1 mapping of the set of subgroups $\neq 1$ onto the set of* *positive integers. If $\mathfrak{Z}$ is finite of order r, then our mapping is 1–1 of the set of subgroups $\neq 1$ onto the set of positive divisors of r that are less than r.*

If $\mathfrak{Z}$ is infinite we can extend our correspondence to include the subgroup 1 consisting of 1 alone by mapping $1 \rightarrow 0$. In the finite case we map $1 \rightarrow r$, so that in all cases we have $\mathfrak{W} = [a^s]$. We note also that in the finite case if $\mathfrak{W} \rightarrow s$, then the order of $\mathfrak{W}$ is $r/s = t$. Hence, we obtain another 1–1 correspondence here by associating with $\mathfrak{W}$ the order of this subgroup. We state this result as

Theorem 4. *Let $\mathfrak{Z}$ be cyclic of order r $(<\infty)$. Then the order of any subgroup of $\mathfrak{Z}$ is a divisor of r and, if t is any positive divisor of r, $\mathfrak{Z}$ possesses one and only one subgroup of order t.*

It is customary to denote the number of positive divisors of an integer r by $d(r)$. Thus $\mathfrak{Z}$ possesses $d(r)$ subgroups.

EXERCISES

1. List the subgroups of the cyclic group of order 12.
2. Let $\mathfrak{Z} = [a]$ be of order $r < \infty$. Show that the order of a^m is $[m,r]/m = r/(m,r)$.
3. Show that a cyclic group of order r possesses exactly $\phi(r)$ generators where $\phi(r)$ (the Euler ϕ-*function*) denotes the number of positive numbers $<r$ that are *prime* to r in the sense $(r,h) = 1$.
4. Show that the subgroup $\mathfrak{H}$ of order t $(r = st)$ of a cyclic group of order r can be characterized in either of the following ways: (1) $\mathfrak{H}$ is the set of sth powers of the elements of $\mathfrak{G}$ or (2) $\mathfrak{H}$ is the set of elements h such that $h^t = 1$.

12. Elementary properties of permutations. A permutation γ which permutes cyclically a set of elements $i_1, i_2, \cdots, i_r$ in the sense that

$$(9) \qquad i_1\gamma = i_2, \quad i_2\gamma = i_3, \quad \cdots, \quad i_{r-1}\gamma = i_r, \quad i_r\gamma = i_1$$

and leaves fixed the other numbers in $\{1, 2, \cdots, n\}$ is called a *cycle.* If γ is of this form, we denote it as $(i_1i_2 \cdots i_r)$. It is clear that we can just as well write

$$\gamma = (i_2i_3 \cdots i_ri_1) = (i_3i_4 \cdots i_ri_1i_2) = \cdots.$$

Two cycles γ and γ' are said to be *disjoint* if their symbols contain no common letters. In this case it is clear that the numbers which are moved by one of these transformations are left fixed by the other. Hence if i is any number and $i\gamma \neq i$, then $i\gamma'\gamma = i\gamma$ and since $i\gamma^2 \neq i\gamma$ also, $i\gamma\gamma' = i\gamma$. Similarly if $i\gamma' \neq i$ then $i\gamma\gamma' = i\gamma'\gamma$ and if $i\gamma = i$ and $i\gamma' = i$ then $i\gamma\gamma' = i\gamma'\gamma$. Thus $\gamma\gamma' = \gamma'\gamma$, that is, any two disjoint cycles commute.

Any permutation α can be written as a product of disjoint cycles. For example, if

$$\alpha = \begin{pmatrix} 1 & 2 & 3 & 4 & 5 & 6 & 7 & 8 \\ 3 & 6 & 5 & 4 & 8 & 2 & 7 & 1 \end{pmatrix},$$

then

$$1\alpha = 3, \quad 3\alpha = 5, \quad 5\alpha = 8, \quad 8\alpha = 1; \quad 2\alpha = 6, \quad 6\alpha = 2;$$

$$4\alpha = 4; \quad 7\alpha = 7;$$

from which it follows that

$$\alpha = (1\ 3\ 5\ 8)(2\ 6)(4)(7).$$

In general, for any α we can begin with any number in $1, 2, \cdots, n$, say i_1, and form $i_1\alpha = i_2$, $i_2\alpha = i_3$, $\cdots$, until we reach a number that occurs previously in this list. The first such repetition occurs when $i_{r+1} = i_r\alpha = i_1$; for $i_k = i_1\alpha^{k-1}$ and if $i_k = i_l$, $l > k$, then $i_1\alpha^{k-1} = i_1\alpha^{l-1}$ and $i_1\alpha^{l-k} = i_1$. Thus the numbers $i_1, i_2, \cdots, i_r$ are permuted cyclically by α. If $r < n$ we can find a j_1 not in this set. If $j_1\alpha^k = i_1\alpha^q$, then $j_1 = i_1\alpha^{q-k}$ is in the original set contrary to assumption. Hence we obtain a new set $\{j_1, j_2, \cdots, j_s\}$ that is cyclically permuted by α and that has no element in common with the original set. If we continue in this way we finally exhaust the set $\{1, 2, \cdots, n\}$. Also it is clear by comparing effects on any number that

$$(10) \qquad \alpha = (i_1 i_2 \cdots i_r)(j_1 j_2 \cdots j_s) \cdots (l_1 l_2 \cdots l_u)$$

where these cycles are disjoint.

A cycle (i) is the identity mapping. Such cycles can be dropped in (10) and we may therefore suppose that $r, s, \cdots, u > 1$ in (10). The factorization thus obtained is unique since we can deduce from it the fact that

$$i_1\alpha = i_2, \quad \cdots, \quad i_{r-1}\alpha = i_r, \quad i_r\alpha = i_1; \quad \cdots; \quad l_1\alpha = l_2, \quad \cdots,$$

$$l_{u-1}\alpha = l_u, \quad l_u\alpha = l_1$$

and that all the other numbers are fixed. If α has the factorization (10) into disjoint cycles, then we shall associate with α the integer

$$(11) \qquad N(\alpha) = (r - 1) + (s - 1) + \cdots + (u - 1).$$

A cycle of the form (ab) is called a *transposition*. It is easy to verify that

$$(12) \qquad (i_1 i_2 \cdots i_r) = (i_1 i_2)(i_1 i_3) \cdots (i_1 i_r).$$

Hence according to (10), α is a product of $N(\alpha)$ transpositions. We shall now show that *if $N(\alpha)$ is even (odd) then any factorization of α as a product of transpositions contains an even (odd) number of factors.* For this purpose we require the following formulas

$$(ac_1 c_2 \cdots c_h b d_1 \cdots d_k)(ab) = (ac_1 \cdots c_h)(bd_1 \cdots d_k)$$

$$(ac_1 \cdots c_h)(bd_1 \cdots d_k)(ab) = (ac_1 \cdots c_h b d_1 \cdots d_k).$$

According to these, if a and b occur in the same cycle in α, then $N(\alpha(ab)) = N(\alpha) - 1$ and, if a and b occur in different cycles in α, then $N(\alpha(ab)) = N(\alpha) + 1$. In any case

$$(13) \qquad N(\alpha(ab)) = N(\alpha) \pm 1.$$

Now suppose that α is a product of m transpositions, say $\alpha = (ab)(cd) \cdots (pq)$. Since $(ab)^{-1} = (ab)$, this implies that

$$\alpha(pq) \cdots (cd)(ab) = 1.$$

Since $N(1) = 0$, iteration of (13) gives

$$0 = N(\alpha) \overset{\overbrace{\hspace{3em}m\hspace{3em}}}{\pm 1 \pm 1 \pm \cdots \pm 1}.$$

Hence $N(\alpha)$ is a sum of m terms $= 1$ or -1. It follows that $N(\alpha)$ is even if and only if m is even. This proves our assertion.

We shall call α *even* or *odd* according as the factorizations of α as a product of transpositions contain even or odd numbers of factors. If α is a product of m transpositions and β is a product of q transpositions, then $\alpha\beta$ is a product of $m + q$ transpositions and α^{-1} is a product of m transpositions. Hence if α is even and β is even, then $\alpha\beta$ is even; if α is even (odd) and β is odd (even), then $\alpha\beta$ is odd and, if both α and β are odd, then $\alpha\beta$ is **even.**

If α is even, then so is α^{-1}. Among other things, these rules show that the set A_n of even permutations is a subgroup of S_n. This subgroup is called *the alternating group*.

EXERCISES

1. Express the elements of S_4 as (1) products of disjoint cycles, (2) products of transpositions. Determine the elements of A_4.

2. Show that, if $n \geq 3$, then any element of A_n is a product of *three-cycles* (abc).

13. Coset decompositions of a group. Suppose first that $\mathfrak{G}$ is an arbitrary transformation group acting in a set S. Then $\mathfrak{G}$ defines an equivalence relation in S by the rule that $x \equiv y(\mathrm{mod}\ \mathfrak{G})$ (read: x is congruent to y modulo $\mathfrak{G}$) if $y = x\alpha$ for some α in $\mathfrak{G}$. That this relation is reflexive, symmetric and transitive is immediate from the definition of a transformation group. It may happen that any two elements of S are equivalent in this sense. In this case $\mathfrak{G}$ is said to be *transitive* in S. In general we obtain a decomposition of S into non-overlapping equivalence classes that we shall call the *transitivity sets* of S relative to $\mathfrak{G}$.

As an instance of this type of decomposition let $S = \{1, 2, \cdots, n\}$ and let $\mathfrak{G} = [\alpha]$ where α is in S_n. If $\alpha = (i_1 i_2 \cdots i_r) \cdots (l_1 l_2 \cdots l_u)$ is the factorization of α into disjoint cycles, then it is clear that $\{i_1, i_2, \cdots, i_r\}, \cdots, \{l_1, l_2, \cdots, l_u\}$ are transitivity sets of $[\alpha]$. The remaining transitivity sets contain single elements. The number $N(\alpha)$ considered in the preceding section can now be defined as $\Sigma(r - 1)$ where r denotes the number of elements in a transitivity set and the sum is taken over these sets. This remark shows again that $N(\alpha)$ is unique and in general it makes somewhat more transparent the discussion of the preceding section.

We suppose now that $\mathfrak{G}$ is any group and that $\mathfrak{H}$ is a subgroup of $\mathfrak{G}$. Let $\mathfrak{H}_r{}'$ be the set of right multiplications in $\mathfrak{G}$ determined by the elements of $\mathfrak{H}$. This means that $\mathfrak{H}_r{}'$ is the set of mappings $x \to xh$, x in $\mathfrak{G}$, h fixed in $\mathfrak{H}$. Since $\mathfrak{H}$ is a subgroup of $\mathfrak{G}$, $\mathfrak{H}_r{}'$ is a subgroup of $\mathfrak{G}_r$; hence $\mathfrak{H}_r{}'$ is a transformation group acting in the set $\mathfrak{G}$. We consider now the transitivity sets determined by $\mathfrak{H}_r{}'$. We write $x \equiv y\ (\mathrm{mod}\ \mathfrak{H})$ in place of $x \equiv y\ (\mathrm{mod}\ \mathfrak{H}_r{}')$. By definition this means that there exists an h in $\mathfrak{H}$ such that $y = xh$, or, equivalently, that $x^{-1}y\ \varepsilon\ \mathfrak{H}$. The transitivity set of elements that are congruent (equivalent) to x is called the *right coset of x relative to $\mathfrak{H}$.*

We shall now introduce a convenient notation for the right cosets. In general, if A and B are subsets of a group $\mathfrak{G}$, then we write AB for the collection of products ab, a in A, b in B. We note that $(AB)C$ is the collection of products $(ab)c$, a in A, b in B, c in C. Since $(ab)c = a(bc)$, any such product is in $A(BC)$. Hence $(AB)C \subseteq A(BC)$. Similarly the reverse inequality holds so that $(AB)C = A(BC)$. The set consisting of a single element x will be denoted as x. Now it is clear that the right coset of x relative to $\mathfrak{H}$ is the set of elements xh, h in $\mathfrak{H}$. Hence this coset is the set $x\mathfrak{H}$. We know, of course, that $\mathfrak{G} = \cup x\mathfrak{H}$ and that either $x\mathfrak{H} = y\mathfrak{H}$ or $x\mathfrak{H} \cap y\mathfrak{H} = \varnothing$.

Examples. (1) Let I_+ be the group of integers relative to addition and let $[m]$ denote the subgroup of multiples of the integer $m > 0$. Here $x \equiv y$ (mod $[m]$) has the same meaning as $x \equiv y$ (mod m) of elementary number theory, namely, $x - y$ is a multiple of m. If x is any integer we can write $x = qm + r$ where $0 \leq r < m$. Then $x \equiv r$ (mod m). Thus any integer is congruent to one of the numbers 0, 1, 2, $\cdots$, $m - 1$. Also it is clear that no two of these numbers are congruent. Hence there are m cosets of I relative to $[m]$:

$$\bar{0} = \{0, \pm m, \pm 2m, \cdots\}$$

$$\bar{1} = \{1, 1 \pm m, 1 \pm 2m, \cdots\}$$

$$\cdots \cdots \cdots \cdots \cdots \cdots$$

$$\overline{(m-1)} = \{m - 1, (m - 1) \pm m, (m - 1) \pm 2m, \cdots\}.$$

(2) $\mathfrak{G} = R_+$, the additive group of real numbers; $\mathfrak{H} = I_+$, the subgroup of integers. Here two real numbers are in the same coset relative to I_+ if and only if their difference is an integer. A coset is therefore a collection of points that are similarly placed in the unit intervals with integer endpoints.

(3) $\mathfrak{G} = S_n$, $\mathfrak{H} = A_n$. If β is even, $\beta \, \varepsilon \, A_n$ and conversely. If β is odd every member of the coset βA_n is odd. Moreover, this coset contains all the odd permutations; for, if γ is odd, then $\beta^{-1}\gamma$ is even and $\gamma \, \varepsilon \, \beta A_n$. Thus we have two cosets: the coset A_n of even permutations and the coset of odd permutations.

Any two right cosets have the same cardinal number, that is, there is a 1–1 correspondence mapping one onto the other. Thus let $x\mathfrak{H}$ and $y\mathfrak{H}$ be arbitrary right cosets and consider the left multiplication $(yx^{-1})_l = x_l^{-1}y_l$. We know that this mapping is 1–1 of $\mathfrak{G}$ onto itself, and it is clear that, if $xh \, \varepsilon \, x\mathfrak{H}$, then $(xh)(yx^{-1})_l = yx^{-1}xh = yh \, \varepsilon \, y\mathfrak{H}$. Hence $(yx^{-1})_l$ induces a 1–1 mapping of $x\mathfrak{H}$ onto $y\mathfrak{H}$. Since $\mathfrak{H} = 1\mathfrak{H}$ is itself a right coset, we see that all the right cosets have the same cardinal number as $\mathfrak{H}$.

We can duplicate the foregoing discussion using left cosets, in place of right cosets. The starting point here is the transformation group $\mathfrak{H}_l' = \{h_l\}$, h in $\mathfrak{H}$. We define the *left congruence relation* relative to the subgroup $\mathfrak{H}$ as the congruence relation determined by the transformation group $\mathfrak{H}_l'$. Thus we set $x \equiv_l y \pmod{\mathfrak{H}}$ for $x \equiv y \pmod{\mathfrak{H}_l'}$. This means simply that there exists an element $h \,\varepsilon\, \mathfrak{H}$ such that $y = hx$, or, equivalently, that $yx^{-1} \,\varepsilon\, \mathfrak{H}$. The equivalence class determined by x is the set $\mathfrak{H}x$ which we shall call the *left coset* of x relative to $\mathfrak{H}$.

One can see by examples (exercise 1 below) that the decomposition of a group into right cosets relative to a subgroup $\mathfrak{H}$ need not coincide with the decomposition into left cosets relative to $\mathfrak{H}$. However, there is a simple relation between these two decompositions, namely, the set of inverses of the elements in any right coset $x\mathfrak{H}$ constitute a left coset. For $(xh)^{-1} = h^{-1}x^{-1} \,\varepsilon\, \mathfrak{H}x^{-1}$ and, as h ranges over $\mathfrak{H}$, $h^{-1}x^{-1}$ ranges over $\mathfrak{H}x^{-1}$. Thus the left coset $\mathfrak{H}x^{-1}$ is uniquely determined by $x\mathfrak{H}$, that is, it does not depend on the element x selected in $x\mathfrak{H}$. It is also immediate that the correspondence $x\mathfrak{H} \to \mathfrak{H}x^{-1}$ is 1–1 of the collection of right cosets onto the collection of left cosets. Hence the collections $\{\mathfrak{H}x\}$ and $\{x\mathfrak{H}\}$ have the same cardinal number. We call this number the *index of $\mathfrak{H}$ in $\mathfrak{G}$*.

Suppose now that $\mathfrak{G}$ is a finite group and that the order of $\mathfrak{G}$ is n. Let $\mathfrak{H}$ be a subgroup of order m and write

$$\mathfrak{G} = a_1\mathfrak{H} \,\cup\, a_2\mathfrak{H} \,\cup \cdots \cup\, a_r\mathfrak{H}$$

where $a_i\mathfrak{H} \cap a_j\mathfrak{H} = \varnothing$ if $i \neq j$. Thus r is the index of $\mathfrak{H}$ in $\mathfrak{G}$. We have seen that each $a_i\mathfrak{H}$ contains m elements. Hence $\mathfrak{G}$ contains mr elements so that $n = mr$. This proves the following fundamental

Theorem 5 (Lagrange). *The order of a subgroup of a finite group is a factor of the order of the group.*

Our result shows that the order of A_n is $n!/2$; for we have seen that the index of A_n in S_n is 2. A second important application of Lagrange's theorem is the

Corollary. *If $\mathfrak{G}$ is a finite group of order n, then $x^n = 1$ for every $x \in \mathfrak{G}$.*

Proof. Let m be the order of $[x]$. Then $x^m = 1$ and $n = mr$. Hence $x^n = 1$.

EXERCISES

1. Determine the coset decompositions of the subgroup $\mathfrak{H} = \{1, (12)\}$ in S_3.

2. Let V be the group of vectors in the plane, vector addition as composition. Show that the vectors that issue from the origin O and have end points on a fixed line through O form a subgroup. What are the cosets relative to this subgroup?

3. Let $\mathfrak{H}_1$ and $\mathfrak{H}_2$ be two subgroups of $\mathfrak{G}$. Show that any coset relative to $\mathfrak{H}_1 \cap \mathfrak{H}_2$ is the intersection of a coset relative to $\mathfrak{H}_1$ with a coset relative to $\mathfrak{H}_2$. Use this result to prove *Poincaré's theorem* that, if $\mathfrak{H}_1$ and $\mathfrak{H}_2$ have finite index in $\mathfrak{G}$, then so has $\mathfrak{H}_1 \cap \mathfrak{H}_2$.

4. Does the rule $x\mathfrak{H} \to \mathfrak{H}x$ define a (single-valued) mapping?

14. Invariant subgroups and factor groups. We wish to determine now the condition on a subgroup $\mathfrak{H}$ in order that we be able to multiply any two congruences modulo $\mathfrak{H}$, that is, that we be able to conclude from any two congruences $x \equiv x'$ (mod $\mathfrak{H}$) and $y \equiv y'$ (mod $\mathfrak{H}$) that $xy \equiv x'y'$ (mod $\mathfrak{H}$). Another way of putting this condition is that, if $x' \in x\mathfrak{H}$ and $y' \in y\mathfrak{H}$, then $x'y' \in xy\mathfrak{H}$. In terms of set multiplication this means that

$$(14) \qquad\qquad (x\mathfrak{H})(y\mathfrak{H}) \subseteq xy\mathfrak{H}$$

holds for all x and y in $\mathfrak{G}$. It is clear that this condition is equivalent to $\mathfrak{H}y\mathfrak{H} \subseteq y\mathfrak{H}$ for all y. Also $\mathfrak{H}y\mathfrak{H} \subseteq y\mathfrak{H}$ implies that $\mathfrak{H}y \subseteq y\mathfrak{H}$. On the other hand, if $\mathfrak{H}$ has this latter property, then

$$\mathfrak{H}y\mathfrak{H} \subseteq y\mathfrak{H}\mathfrak{H} = y\mathfrak{H}$$

since $\mathfrak{H}^2 = \mathfrak{H}$. It is clear also that the condition $\mathfrak{H}y \subseteq y\mathfrak{H}$ is equivalent to $y^{-1}\mathfrak{H}y \subseteq \mathfrak{H}$, and we use this form of the condition in the following

Definition 4. *A subgroup $\mathfrak{H}$ is called* invariant (normal, self-conjugate, distinguished) *if $y^{-1}\mathfrak{H}y \subseteq \mathfrak{H}$ for every y in $\mathfrak{G}$.*

Our remarks show that $\mathfrak{H}$ is invariant if and only if $(x\mathfrak{H})(y\mathfrak{H}) \subseteq xy\mathfrak{H}$ holds for every x, y in $\mathfrak{G}$. In terms of elements the test for invariance of a subgroup $\mathfrak{H}$ is that, if $h \in \mathfrak{H}$ and y is arbitrary, then

$y^{-1}hy \ \varepsilon \ \mathfrak{H}$. Since $\mathfrak{H}y \subseteq y\mathfrak{H}$ for all y, $\mathfrak{H}y^{-1} \subseteq y^{-1}\mathfrak{H}$ and multiplication by y on the right and on the left gives $y\mathfrak{H} \subseteq \mathfrak{H}y$. Hence $\mathfrak{H}y = y\mathfrak{H}$. Thus if $\mathfrak{H}$ is invariant, then the right coset determined by any element coincides with the left coset determined by this element. Hence there is only one coset decomposition for an invariant subgroup.

If $\mathfrak{H}$ is invariant, then $(x\mathfrak{H})(y\mathfrak{H}) = x\mathfrak{H}y\mathfrak{H} = xy\mathfrak{H}\mathfrak{H} = xy\mathfrak{H}$. Hence the set of cosets of $\mathfrak{H}$ is closed relative to set multiplication. We now show that the collection $\mathfrak{G}/\mathfrak{H}$ of cosets and this composition is a group. The associative law holds for this composition since multiplication of sets is associative. The coset $\mathfrak{H}$ acts as the identity since $\mathfrak{H}(x\mathfrak{H}) = x\mathfrak{H}$ and $(x\mathfrak{H})\mathfrak{H} = x\mathfrak{H}$. Also $x\mathfrak{H}$ has the inverse $x^{-1}\mathfrak{H}$ since $(x\mathfrak{H})(x^{-1}\mathfrak{H}) = \mathfrak{H} = (x^{-1}\mathfrak{H})(x\mathfrak{H})$. This proves our assertion. The group consisting of the set of cosets and the composition that we have defined is called the *factor (quotient) group* $\mathfrak{G}/\mathfrak{H}$ of $\mathfrak{G}$ relative to the invariant subgroup $\mathfrak{H}$. Clearly the order of $\mathfrak{G}/\mathfrak{H}$ is the index of $\mathfrak{H}$ in $\mathfrak{G}$.

Examples. (1) I, the group of integers relative to addition; $[m]$, the subgroup of multiples of the integer $m > 1$. $[m]$ is invariant since it is clear that any subgroup of a commutative group is invariant. The factor group $I/[m]$ is cyclic with $\bar{1} = 1 + [m]$ as generator. (2) A_n is an invariant subgroup of S_n. For if α is even $\beta^{-1}\alpha\beta$ is even for any β. The factor group S_n/A_n has order 2.

EXERCISES

1. Prove that any subgroup of index 2 is invariant.
2. Show that $\mathfrak{H} = \{1, (1\ 2)\}$ is not invariant in S_3.
3. Show that the subgroup of transformations of the form $x \rightarrow x + b$ is invariant in the group of transformations $x \rightarrow ax + b$, $a \neq 0$.

15. Homomorphism of groups. The concepts of isomorphism and of isomorphic groups become considerably more fruitful when they are generalized in the manner that we shall now indicate. The generalizations that we wish to define are obtained by dropping the requirement of 1–1 ness in our previous definitions. Thus we have the following fundamental

Definition 5. *A mapping η of a group $\mathfrak{G}$ into a group $\mathfrak{G}'$ is called a* homomorphism *if $(xy)\eta = (x\eta)(y\eta)$. If η is a homomorphism of $\mathfrak{G}$ onto $\mathfrak{G}'$, then $\mathfrak{G}'$ is called a* homomorphic image *of $\mathfrak{G}$.*

An important instance of a homomorphism is obtained by taking a factor group $\mathfrak{G}/\mathfrak{H}$ of $\mathfrak{G}$ relative to an invariant subgroup $\mathfrak{H}$ of $\mathfrak{G}$. By definition $(x\mathfrak{H})(y\mathfrak{H}) = (xy)\mathfrak{H}$ in $\mathfrak{G}/\mathfrak{H}$. Hence if we map the element x of $\mathfrak{G}$ into its coset $x\mathfrak{H}$, then we obtain a homomorphism of $\mathfrak{G}$ onto $\mathfrak{G}/\mathfrak{H}$. Thus any factor group of $\mathfrak{G}$ is a homomorphic image of $\mathfrak{G}$.

It should be noted that the definition that we have given does not require that η be a mapping onto $\mathfrak{G}$. If η is 1–1, then we call it an *isomorphism of $\mathfrak{G}$ into $\mathfrak{G}'$*. Previously we have dealt exclusively with *isomorphisms onto* and with *isomorphic groups*. We consider now some concrete examples of homomorphisms.

Examples. (1) Let $\mathfrak{G} = R_+$, the additive group of real numbers, and let $\mathfrak{G}' = U$, the multiplicative group of complex numbers of absolute value 1. The mapping $\theta \to e^{i\theta}$ is a homomorphism of $\mathfrak{G}$ onto $\mathfrak{G}'$ since $e^{i(\theta_1+\theta_2)} = e^{i\theta_1}e^{i\theta_2}$ and every element of $\mathfrak{G}'$ has the form $e^{i\theta}$. This mapping is not an isomorphism and, in fact, it is easy to see that these groups are not isomorphic (exercise 3 below).

(2) Let $\mathfrak{G} = V$ the group of plane vectors (α,β) with the usual composition $(\alpha,\beta) + (\alpha',\beta') = (\alpha + \alpha',\ \beta + \beta')$. The mapping $(\alpha,\beta) \to \alpha$ is a homomorphism of V onto R_+.

(3) Let $\mathfrak{G}$ be the symmetric group S_n and map the permutation $\tau \,\varepsilon\, S_n$ on the number 1 or on the number -1 according as τ is even or odd. In any case denote the image as $\chi(\tau)$. Then $\chi(\tau\tau') = \chi(\tau)\chi(\tau')$. Hence $\tau \to \chi(\tau)$ is a homomorphism of S_n onto the multiplicative group of numbers 1, -1.

(4) Consider the additive group of integers I_+ and any group $\mathfrak{G}$. Let a be a definite element of $\mathfrak{G}$. Then the mapping $n \to a^n$, n in I_+, satisfies $a^{m+n} = a^m a^n$. Hence it is a homomorphism of I_+ into $\mathfrak{G}$.

We derive next some of the elementary properties of homomorphisms. We note first the following

Theorem 6. *The image $\mathfrak{G}\eta$ of a homomorphism of $\mathfrak{G}$ into $\mathfrak{G}'$ is a subgroup of $\mathfrak{G}'$.*

Proof. Since $(x\eta)(y\eta) = (xy)\eta$, $\mathfrak{G}\eta$ is closed under the composition in $\mathfrak{G}'$. Also $(1\eta)(1\eta) = 1\eta$ so that 1η is the identity $1'$ of $\mathfrak{G}'$. Finally $(x\eta)(x^{-1}\eta) = 1\eta = 1'$, and this means that $(x\eta)^{-1} = x^{-1}\eta$ is in $\mathfrak{G}\eta$.

We consider next the totality $\mathfrak{K}$ of elements k of $\mathfrak{G}$ such that $k\eta = 1'$. This is the inverse image set $\eta^{-1}(1')$ of the identity element $1'$ of $\mathfrak{G}'$. Since $1\eta = 1'$, $\mathfrak{K} \ni 1$. Hence if $\mathfrak{K} \neq 1$, then η is not 1–1. On the other hand, we shall show now that, if $\mathfrak{K} = 1$, *then η is an isomorphism*. Thus assume that $a\eta = b\eta$. Then

$(a^{-1}b)\eta = a^{-1}\eta b\eta = (a\eta)^{-1}(b\eta) = 1'$. Hence $a^{-1}b = 1$ and $a = b$. We prove next

Theorem 7. *If η is a homomorphism of $\mathfrak{G}$ into $\mathfrak{G}'$, the inverse image $\mathfrak{K} = \eta^{-1}(1')$ of the identity of $\mathfrak{G}'$ is an invariant subgroup of $\mathfrak{G}$.*

Proof. We know that $1 \, \varepsilon \, \mathfrak{K}$. If k_1, $k_2 \, \varepsilon \, \mathfrak{K}$, then $(k_1 k_2)\eta = (k_1\eta)(k_2\eta) = 1'1' = 1'$. Hence $k_1 k_2 \, \varepsilon \, \mathfrak{K}$. Also if $k \, \varepsilon \, \mathfrak{K}$, then $k^{-1}\eta = (k\eta)^{-1} = 1'^{-1} = 1'$ and $k^{-1} \, \varepsilon \, \mathfrak{K}$. This proves that $\mathfrak{K}$ is a subgroup. Finally if a is arbitrary in $\mathfrak{G}$ and $k \, \varepsilon \, \mathfrak{K}$, then $(a^{-1}ka)\eta = (a^{-1}\eta)(k\eta)(a\eta) = (a\eta)^{-1}1'(a\eta) = 1'$ so that $a^{-1}ka \, \varepsilon \, \mathfrak{K}$. Hence $\mathfrak{K}$ is invariant.

The group $\mathfrak{K} = \eta^{-1}(1')$ is called *the kernel* of the homomorphism η.

EXERCISES

1. Determine the kernels of the homomorphisms in the foregoing examples.

2. Prove the following extension of Theorem 6: Let $\mathfrak{G}$ be a group and let $\mathfrak{G}'$ be any set in which a composition $a'b'$ is defined. Suppose that η is a mapping of $\mathfrak{G}$ into $\mathfrak{G}'$ such that $(xy)\eta = (x\eta)(y\eta)$. Then the image $\mathfrak{G}\eta$ is a group relative to the composition defined in $\mathfrak{G}'$.

3. Prove that the groups R_+ and U of example 1 are not isomorphic.

4. Let $\mathfrak{G}$ be the transformation group of mappings $x \rightarrow ax + b$ where a and b are real numbers and $a \neq 0$. Show that the correspondence that associates with the indicated transformation the real number a is a homomorphism of $\mathfrak{G}$ onto R^*. What is the kernel?

5. Show that if k is an integer then the mapping $e^{i\theta} \rightarrow e^{ki\theta}$ is a homomorphism of U onto itself. Determine the kernel.

16. The fundamental theorem of homomorphism for groups. We have seen that the mapping $x \rightarrow \bar{x} = x\mathfrak{H}$ is a homomorphism of the group $\mathfrak{G}$ onto its factor group $\bar{\mathfrak{G}} = \mathfrak{G}/\mathfrak{H}$ relative to the invariant subgroup $\mathfrak{H}$. We shall call this homomorphism *the natural homomorphism* of $\mathfrak{G}$ onto $\bar{\mathfrak{G}}$ and in the sequel we denote it by the letter ν. The kernel of ν, that is, the set of elements a such that $a\nu \equiv a\mathfrak{H} = \mathfrak{H}$ is obviously the given invariant subgroup $\mathfrak{H}$.

We note next that, if η is a homomorphism of $\mathfrak{G}$ into $\mathfrak{G}'$ and ρ is a homomorphism of $\mathfrak{G}'$ into $\mathfrak{G}''$, then $\eta\rho$ is a homomorphism of $\mathfrak{G}$ into $\mathfrak{G}''$. This is immediate from the definition. In particular we see that, if ν is the natural homomorphism of $\mathfrak{G}$ onto $\bar{\mathfrak{G}} = \mathfrak{G}/\mathfrak{H}$

and $\bar{\eta}$ is a homomorphism of $\mathfrak{G}$ into another group $\mathfrak{G}'$, then the resultant $\nu\bar{\eta}$ is a homomorphism of $\mathfrak{G}$ into $\mathfrak{G}'$. The kernel of this homomorphism evidently contains $\mathfrak{H}$.

Conversely let η be a homomorphism of $\mathfrak{G}$ into a second group $\mathfrak{G}'$ and let $\mathfrak{H}$ be an invariant subgroup of $\mathfrak{G}$ contained in the kernel $\mathfrak{K} = \eta^{-1}(1')$. Let a and b be two elements in the same coset relative to $\mathfrak{H}$. Then $b = ah$, h in $\mathfrak{H}$, and $b\eta = (a\eta)(h\eta) = (a\eta)1' = a\eta$. This shows that the rule $a\mathfrak{H} \rightarrow a\eta$ defines a single-valued mapping of $\bar{\mathfrak{G}} = \mathfrak{G}/\mathfrak{H}$ into $\mathfrak{G}'$. We denote this mapping as $\bar{\eta}$ and we prove that it is a homomorphism. This follows from

$$[(a\mathfrak{H})(b\mathfrak{H})]\bar{\eta} = (ab\mathfrak{H})\bar{\eta} = (ab)\eta = (a\eta)(b\eta) = ((a\mathfrak{H})\bar{\eta})((b\mathfrak{H})\bar{\eta}).$$

We shall call $\bar{\eta}$ the induced homomorphism of $\bar{\mathfrak{G}}$ into $\mathfrak{G}'$. Evidently $a\nu\bar{\eta} = (a\mathfrak{H})\bar{\eta} = a\eta$ so that the given homomorphism permits the factorization $\eta = \nu\bar{\eta}$.

We note next that, if $(a\mathfrak{H})\bar{\eta} = 1'$, then $a\eta = 1'$ and $a \in \mathfrak{K}$. Also the converse holds. Hence we see that the kernel of $\bar{\eta}$ is the totality $\mathfrak{K}/\mathfrak{H}$ of cosets of the form $k\mathfrak{H}$, k in $\mathfrak{K}$. As a consequence we see that $\bar{\eta}$ is 1–1 if and only if $\mathfrak{K} = \mathfrak{H}$. This completes the proof of the important

Theorem 8. *Let η be a homomorphism of $\mathfrak{G}$ into $\mathfrak{G}'$ and let $\mathfrak{H}$ be an invariant subgroup of $\mathfrak{G}$ contained in $\mathfrak{K} = \eta^{-1}(1')$. Then the rule $a\mathfrak{H} \rightarrow a\eta$ is a homomorphism $\bar{\eta}$ of $\bar{\mathfrak{G}} = \mathfrak{G}/\mathfrak{H}$ into $\mathfrak{G}'$ and $\eta = \nu\bar{\eta}$ where ν is the natural homomorphism of $\mathfrak{G}$ onto $\bar{\mathfrak{G}}$. The mapping $\bar{\eta}$ is an isomorphism if and only if $\mathfrak{K} = \mathfrak{H}$.*

Suppose now that we particularize our considerations to the case in which η is a homomorphism of $\mathfrak{G}$ onto $\mathfrak{G}'$. If $\mathfrak{K}$ is the kernel, then we see that the induced mapping $\bar{\eta}$ of $\bar{\mathfrak{G}} = \mathfrak{G}/\mathfrak{K}$ onto $\mathfrak{G}'$ is an isomorphism. Hence $\bar{\mathfrak{G}} \cong \mathfrak{G}'$. This, together with the result noted in the first paragraph, proves the

Fundamental theorem of homomorphism for groups. *Any factor group of $\mathfrak{G}$ is a homomorphic image of $\mathfrak{G}$ and conversely if $\mathfrak{G}'$ is a homomorphic image of $\mathfrak{G}$ then $\mathfrak{G}'$ is isomorphic to a factor group of $\mathfrak{G}$.*

As an illustration of the power of this theorem we use it to derive again a part of the theory of cyclic groups. Let $\mathfrak{G} = [a]$

be cyclic with generator a. Then we know that the mapping $n \to a^n$ is a homomorphism of I_+ onto $\mathfrak{G}$. Hence $\mathfrak{G} \cong I_+/\mathfrak{H}$ where $\mathfrak{H}$, the kernel, is a subgroup of I_+. Now we use our determination of the subgroups of I_+. According to this we have either $\mathfrak{H} = 0$ or $\mathfrak{H} = [m]$ where $m > 0$. In the former case the mapping $n \to a^n$ is an isomorphism, and $\mathfrak{G} \cong I_+$. Otherwise we see that $\mathfrak{G} \cong I_+/[m]$, a group of order m. It is immediate from these remarks that any two cyclic groups of the same order are isomorphic.

EXERCISES

1. Prove that $R_+/[2\pi] \cong U$ where R_+ and U are as in example 1 of p. 42 and $[2\pi]$ is the cyclic group generated by 2π.

2. Let $[x]$ be a cyclic group of order s, and $[y]$ a cyclic group of order t. Show that there is a homomorphism η of $[x]$ into $[y]$, such that $x\eta = y^k$, if and only if sk is a multiple of t. If $sk = mt$, show that η is an isomorphism if and only if $(s,m) = 1$.

17. Endomorphisms, automorphisms, center of a group. A homomorphism of a group *into* itself is called an *endomorphism*; an isomorphism of a group *onto* itself is called an *automorphism*. The resultant of endomorphisms is an endomorphism. Hence the set $\mathfrak{E}$ of endomorphisms of a group $\mathfrak{G}$ is a sub-semi-group of the semi-group of single-valued mappings in the set $\mathfrak{G}$. Evidently the identity mapping is an endomorphism; hence the semi-group $\mathfrak{E}$ has an identity.

Consider next the set $\mathfrak{A}$ of automorphisms of the group $\mathfrak{G}$. We assert that $\mathfrak{A}$ is the group of units of $\mathfrak{E}$. For if α is a unit in $\mathfrak{E}$, α^{-1} exists and hence α is 1–1 of $\mathfrak{G}$ onto itself. On the other hand, if α is an automorphism, its inverse α^{-1} is also an automorphism; for

$$(xy)\alpha^{-1} = ((x\alpha^{-1}\alpha)(y\alpha^{-1}\alpha))\alpha^{-1} = (((x\alpha^{-1})(y\alpha^{-1}))\alpha)\alpha^{-1}$$

$$= (x\alpha^{-1})(y\alpha^{-1}).$$

Hence α has an inverse in $\mathfrak{E}$. This proves also that $\mathfrak{A}$ is a group of transformations in $\mathfrak{G}$. We shall call this group the *group of automorphisms* of $\mathfrak{G}$.

If a is a fixed element the mapping

$$(15) \qquad\qquad C_a: \quad x \to a^{-1}xa$$

is an automorphism of $\mathfrak{G}$, since

$$a^{-1}(xy)a = (a^{-1}xa)(a^{-1}ya)$$

and, as is easy to verify, C_a is 1–1 of $\mathfrak{G}$ onto itself. As a matter of fact, the 1–1 ness is clear if we note that

$$(16) \qquad C_a = a_r a_l^{-1} = a_l^{-1} a_r$$

where, as usual, a_r and a_l are respectively the right and the left multiplications determined by a. The automorphism C_a is called the *inner automorphism* determined by the element a.

We shall now show that the set $\mathfrak{J}$ of inner automorphisms forms an invariant subgroup of the group of automorphisms $\mathfrak{A}$. Let C_{a_1} and C_{a_2} be inner automorphisms. Then

$$xC_{a_1}C_{a_2} = a_2^{-1}a_1^{-1}xa_1a_2 = (a_1a_2)^{-1}x(a_1a_2) = xC_{a_1a_2}$$

so that

$$(17) \qquad C_{a_1a_2} = C_{a_1}C_{a_2}.$$

This equation shows that the correspondence $a \to C_a$ is a homomorphism of $\mathfrak{G}$ into its group of automorphisms. It follows (Theorem 6) that the image set $\mathfrak{J}$ is a subgroup of $\mathfrak{A}$. Now let α be any automorphism and consider the product $\alpha^{-1}C_a\alpha$. Since

$$x\alpha^{-1}C_a\alpha = (a^{-1}(x\alpha^{-1})a)\alpha = (a^{-1}\alpha)x(a\alpha)$$

$$= (a\alpha)^{-1}x(a\alpha)$$

$$= xC_{a\alpha},$$

$$(18) \qquad \alpha^{-1}C_a\alpha = C_{a\alpha}$$

is inner. This proves the invariance of $\mathfrak{J}$. The factor group $\mathfrak{A}/\mathfrak{J}$ is called the *group of outer automorphisms* of the group $\mathfrak{G}$.

We return to the homomorphism $a \to C_a$ of $\mathfrak{G}$ onto $\mathfrak{J}$. The kernel $\mathfrak{C}$ of this mapping is the set of elements c such that $C_c = 1$. Thus $c \, \varepsilon \, \mathfrak{C}$ if and only if $c^{-1}xc = x$ for all x or equivalently,

$$(19) \qquad cx = xc$$

for all x. We shall call $\mathfrak{C}$ the *center* of the group $\mathfrak{G}$. By Theorem 7 or directly we see that $\mathfrak{C}$ is an invariant subgroup. Also by the fundamental theorem of homomorphism $\mathfrak{J} \cong \mathfrak{G}/\mathfrak{C}$. We summarize our results in the following

Theorem 9. *The set $\mathfrak{J}$ of inner automorphisms is an invariant subgroup of the group of automorphisms and $\mathfrak{J} \cong \mathfrak{G}/\mathfrak{C}$ where $\mathfrak{C}$ is the center of the group.*

EXERCISES

1. Prove that the mapping $a \to a^{-1}$ is an automorphism if and only if $\mathfrak{G}$ is commutative.

2. Show that, if k is an integer and $\mathfrak{G}$ is commutative, then $a \to a^k$ is an endomorphism.

3. Determine the group of automorphisms of any cyclic group.

4. Determine the group of automorphisms of the symmetric group S_3.

5. The transformation group generated by the group of automorphisms and the group of right multiplications is called the *holomorph* $\mathfrak{H}$ of the group $\mathfrak{G}$. Show that (1) $\mathfrak{H}$ contains all the left multiplications, (2) any element of $\mathfrak{H}$ can be written in one and only one way as a product αa_r of an automorphism α and a right multiplication a_r, (3) if $\mathfrak{G}$ is finite, then the order of $\mathfrak{H}$ is the product of the order of $\mathfrak{G}$ by the order of $\mathfrak{A}$.

18. Conjugate classes. The elements x and y of $\mathfrak{G}$ are said to be *conjugate* if they are equivalent relative to the congruence relation determined by the transformation group $\mathfrak{J}$. This means that there exists an a in $\mathfrak{G}$ such that $a^{-1}xa = y$. The transitivity sets determined by the group $\mathfrak{J}$ are called the *conjugate classes* of the group $\mathfrak{G}$. The conjugate class determined by the element c consists of a single element if and only if c is in the center of the group.

As an illustration of these ideas we shall determine the conjugate classes of the symmetric group S_n. We remark first that if α is the permutation

$$\begin{pmatrix} 1 & 2 & \cdots & n \\ 1\alpha & 2\alpha & \cdots & n\alpha \end{pmatrix}$$

and β is arbitrary, then $\beta^{-1}\alpha\beta$ sends 1β into $1\alpha\beta$ so that $\beta^{-1}\alpha\beta$ can be represented by the symbol

$$\begin{pmatrix} 1\beta & 2\beta & \cdots & n\beta \\ 1\alpha\beta & 2\alpha\beta & \cdots & n\alpha\beta \end{pmatrix}.$$

It follows that if

$$(20) \qquad \alpha = (i_1 i_2 \cdots i_r)(j_1 j_2 \cdots j_s) \cdots (l_1 l_2 \cdots l_u)$$

then

$$(21) \qquad \beta^{-1}\alpha\beta = (i_1\beta\, i_2\beta \cdots i_r\beta) \cdots (l_1\beta\, l_2\beta \cdots l_u\beta).$$

We may suppose that $r \geq s \geq \cdots \geq u$ and that all the numbers are displayed in (20). Then $r + s + \cdots + u = n$. In this way we associate with α a set of positive integers $r, s, \cdots, u$ such that

$$(22) \qquad r \geq s \geq \cdots \geq u, \quad r + s + \cdots + u = n.$$

Equation (21) shows that α and α' are conjugates in S_n if and only if the associated sets $r, s, \cdots, u$ are the same for these two permutations. A system of integers satisfying (22) is called a *partition* of n. Hence we have a 1–1 correspondence between the conjugate classes in S_n and the partitions of n. The number of conjugate classes coincides with the number $p(n)$ of distinct partitions of n. The function $p(n)$ is an important arithmetic function. Its first few values are

$$p(2) = 2, \quad p(3) = 3, \quad p(4) = 5, \quad p(5) = 7, \quad p(6) = 11.$$

Also it is clear from (21) that, if $r > 1$ and $n > 2$, then β can be chosen so that $\beta^{-1}\alpha\beta \neq \alpha$. Hence, if $\alpha \neq 1$, then there exists a β such that $\beta\alpha \neq \alpha\beta$. This shows that the center of S_n, $n > 2$, is the identity.

EXERCISES

1. Prove that, if $\mathfrak{G}$ is a finite permutation group, then the number of elements in any transitivity set determined by $\mathfrak{G}$ is a factor of the order of the group. (Hint: If i is any number in the set $S = \{1, 2, \cdots, n\}$, the set of transformations $\alpha \, \varepsilon \, \mathfrak{G}$ that leave i fixed is a subgroup $\mathfrak{H}$. Show that the elements in the transitivity set containing i can be put into 1–1 correspondence with the left cosets of $\mathfrak{H}$. Hence prove that the number of elements in the transitivity set is the index of $\mathfrak{H}$ in $\mathfrak{G}$.)

2. Prove that the number of elements in any conjugate class of a finite group $\mathfrak{G}$ is a factor of the order of $\mathfrak{G}$.

3. Prove that the center of a group of prime power order contains more than one element.

Chapter II

RINGS, INTEGRAL DOMAINS AND FIELDS

In this chapter we begin the study of a second important type of algebraic system called a *ring*. As we shall see, rings are sets with two suitably restricted binary compositions. Unlike the theory of groups which had essentially one source, namely, the study of sets of 1–1 transformations relative to resultant composition, the theory of rings has been fused out of a number of special theories. For this reason it will appear to be somewhat less unified than the theory of groups. In the present chapter we introduce the basic concepts of integral domain, division ring, field, ideal, difference ring, isomorphism, homomorphism and anti-isomorphism. Also we introduce some important special instances of rings such as matrix rings and quaternions. Finally we prove the analogue for rings of Cayley's theorem on groups.

1. Definition and examples.

Definition 1. *A* ring *is a system consisting of a set $\mathfrak{A}$ and two binary compositions in $\mathfrak{A}$ called* addition *and* multiplication *such that*

1. *$\mathfrak{A}$ together with addition $(+)$ is a commutative group.*
2. *$\mathfrak{A}$ together with multiplication $(\cdot)$ is a semi-group.*
3. *The distributive laws*

$$\text{D} \qquad a(b + c) = ab + ac$$

$$(b + c)a = ba + ca$$

hold.

49

Thus the assumptions included under 1 and 2 are that $a + b$ and $ab \,\varepsilon\, \mathfrak{A}$ and satisfy the following conditions:

A1 $$(a + b) + c = a + (b + c).$$

A2 $$a + b = b + a.$$

A3   There is an element 0 such that $a + 0 = a = 0 + a$.

A4   For each a there is a negative $-a$ such that $a + (-a) = 0 = -a + a$.

M $$(ab)c = a(bc).$$

The system $\mathfrak{A}, +$ will be called *the additive group* and the system $\mathfrak{A}, \cdot$ will be called *the multiplicative semi-group* of the ring.

Examples. (1) The set I of integers with the ordinary addition and multiplication operations. We have noted in the Introduction that this is a ring.

(2) The set R_0 of rational numbers with the usual addition and multiplication. A rigorous definition of this ring will be given in the next chapter.

(3) The set R of real numbers with the usual addition and multiplication.

(4) The set $I[\sqrt{2}\,]$ of real numbers of the form $m + n\sqrt{2}$ where m and n are integers, addition and multiplication as usual. Clearly the sum and difference of two numbers in $I[\sqrt{2}\,]$ belong to this set. Also

$$(m + n\sqrt{2}\,)(m' + n'\sqrt{2}\,) = (mm' + 2nn') + (mn' + nm')\sqrt{2}$$

so that $I[\sqrt{2}\,]$ is closed under multiplication. It follows easily that this system is a ring (see the discussion of subrings in § 5).

(5) The set $R_0[\sqrt{2}\,]$ of real numbers of the form $a + b\sqrt{2}$ where a and b are rational numbers, addition and multiplication as usual.

(6) The set C of complex numbers with the usual addition and multiplication.

(7) The set $I[\sqrt{-1}\,]$ of complex numbers of the form $m + n\sqrt{-1}$, m and n integers with ordinary addition and multiplication. This example is similar to (4).

(8) The set Γ of real valued continuous functions on the interval $[0,1]$ where $(f + g)(x) = f(x) + g(x)$ and $(fg)(x) = f(x)g(x)$.

(9) The set consisting of the two elements 0, 1 with the following addition and multiplication tables:

+	0	1
0	0	1
1	1	0

$\cdot$	0	1
0	0	0
1	0	1

EXERCISES

1. Let A be the set of all real valued functions on $(-\infty, \infty)$. Show that A is a group with the ordinary addition and that A is a semi-group relative to $f \cdot g(x) = f(g(x))$. Is A a ring relative to these two compositions?

2. Show that the three elements 0, 1, 2 constitute a ring if addition and multiplication are defined by the following tables

<table>
<tr><td colspan="4" align="center">+</td><td></td><td colspan="4"></td></tr>
<tr><td></td><td>0</td><td>1</td><td>2</td><td></td><td></td><td>0</td><td>1</td><td>2</td></tr>
<tr><td>0</td><td>0</td><td>1</td><td>2</td><td></td><td>0</td><td>0</td><td>0</td><td>0</td></tr>
<tr><td>1</td><td>1</td><td>2</td><td>0</td><td></td><td>1</td><td>0</td><td>1</td><td>2</td></tr>
<tr><td>2</td><td>2</td><td>0</td><td>1</td><td></td><td>2</td><td>0</td><td>2</td><td>1</td></tr>
</table>

A number of elementary properties of rings are consequences of the fact that a ring is a group relative to addition and a semi-group relative to multiplication. For example, we have $-(a + b) = -a - b \equiv -a + (-b)$ and, if na is defined for the integer n as before, then the rules for multiples

$$n(a + b) = na + nb$$

$$(n + m)a = na + ma$$

$$(nm)a = n(ma)$$

hold. Also the generalized associative laws hold for addition and for multiplication, and the generalized commutative law holds for addition. There are also a number of other simple results that follow from the distributive laws. In the first place, induction on m and n gives the generalization

$$(a_1 + a_2 + \cdots + a_m)(b_1 + b_2 + \cdots + b_n)$$

$$= a_1 b_1 + a_1 b_2 + \cdots + a_1 b_n + a_2 b_1 + a_2 b_2 + \cdots + a_2 b_n + \cdots$$

$$+ a_m b_1 + \cdots + a_m b_n,$$

or

$$\left(\sum_1^m a_i \right) \left(\sum_1^n b_j \right) = \sum_{i=1, j=1}^{m,n} a_i b_j.$$

We note next that

$$a0 = 0 = 0a$$

for all a; for we have $a0 = a(0 + 0) = a0 + a0$. Addition of $-a0$ gives $a0 = 0$. Similarly $0a = 0$. We have the equation

$$0 = 0b = (a + (-a))b = ab + (-a)b,$$

which shows that

$$(-a)b = -ab.$$

Similarly $a(-b) = -ab$ and consequently

$$(-a)(-b) = -a(-b) = -(-ab) = ab.$$

EXERCISES

1. Prove that $a(b - c) = ab - ac$.

2. Prove that for any integer n, $n(ab) = (na)b = a(nb)$.

3. Let $\mathfrak{A}$ be a system which satisfies all the conditions for a ring except commutativity of addition. Prove that, if $\mathfrak{A}$ contains an element c that can be *left cancelled* in the sense that $ca = cb$ implies $a = b$, then $\mathfrak{A}$ is a ring.

If a and b commute in the sense that $ab = ba$, then the powers of a commute with the powers of b and we can prove by induction the important *binomial theorem:*

$$(1) \qquad (a + b)^n = a^n + \binom{n}{1} a^{n-1}b + \binom{n}{2} a^{n-2}b^2 + \cdots + b^n,$$

where $\binom{n}{i}$ is an integer and is given by the formula

$$(2) \qquad \binom{n}{i} = \frac{n!}{i!(n - i)!}.$$

This is evident if $n = 1$. Assume now that

$$(3) \qquad (a + b)^r = \sum_{k=0}^{r} \binom{r}{k} a^k b^{r-k}.$$

We use here the convention that $0! = 1$ so that (3) agrees with (1) for $n = r$. Now multiply both sides of (3) by $a + b$. Then we obtain

$$(a + b)^{r+1} = \sum_{k=0}^{r} \binom{r}{k} a^{k+1}b^{r-k} + \sum_{k=0}^{r} \binom{r}{k} a^k b^{r-k+1}.$$

The term $a^k b^{r+1-k}$, $k \neq 0, r + 1$, in the right-hand side of this equation has the coefficient

$$\binom{r}{k} + \binom{r}{k-1} = \frac{r!}{k!(r-k)!} + \frac{r!}{(k-1)!(r-k+1)!}$$

$$= \frac{r!(r-k+1) + r!k}{k!(r-k+1)!}$$

$$= \frac{(r+1)!}{k!(r-k+1)!} = \binom{r+1}{k}.$$

Hence (1) holds for $n = r + 1$ and this completes the proof.

2. Types of rings. We obtain various types of rings by imposing conditions on the multiplicative semi-group. Thus a ring $\mathfrak{A}$ is said to be *commutative* if its multiplicative semi-group is commutative. The ring $\mathfrak{A}$ is said to have an *identity* if its multiplicative semi-group has an identity. If such an element exists, it is unique. All of the examples listed above are commutative and have identities. An example of a ring without an identity is the set of even integers. Examples of non-commutative rings will be given in §§ 4–5. If the identity $1 = 0$, any $a = a1 = a0 = 0$ so that $\mathfrak{A}$ has only one element. In other words, if $\mathfrak{A} \neq 0$, then $1 \neq 0$.

A ring is called an *integral domain* (*domain of integrity*) if the set $\mathfrak{A}^*$ of non-zero elements determines a sub-semi-group of the multiplicative semi-group. This, of course, means simply that, if $a \neq 0$ and $b \neq 0$ in $\mathfrak{A}$, then $ab \neq 0$. All of the foregoing examples except (8) are of this type. On the other hand, in (8) we can take the two elements

$$f(x) = \begin{cases} 0 \text{ for } 0 \leq x \leq \tfrac{1}{2} \\ x - \tfrac{1}{2} \text{ for } \tfrac{1}{2} < x \leq 1 \end{cases}$$

$$g(x) = \begin{cases} -x + \tfrac{1}{2} \text{ for } 0 \leq x \leq \tfrac{1}{2} \\ 0 \text{ for } \tfrac{1}{2} < x \leq 1 \end{cases}.$$

Then $f \neq 0$ (the constant function 0) and $g \neq 0$ but $fg = 0$. Hence the ring of continuous functions on [0,1] is not an integral domain.

If a is an element of a ring $\mathfrak{A}$ for which there exists a $b \neq 0$ such that $ab = 0$ ($ba = 0$), then a is called a *left* (*right*) *zero-divisor* in $\mathfrak{A}$. Clearly the element 0 is a left and right zero-divisor if $\mathfrak{A}$ contains more than one element. If $a \neq 0$ is a left zero-divisor and $ab = 0$ for $b \neq 0$, then b is a right zero-divisor $\neq 0$. It is therefore clear from the definitions that a ring is an integral domain if and only if it possesses no zero-divisors $\neq 0$.

We note also that a ring is an integral domain if and only if the restricted cancellation laws of multiplication hold, that is, $ab = ac$, $a \neq 0$ imply $b = c$ and $ba = ca$, $a \neq 0$ imply $b = c$. Thus assume that $\mathfrak{A}$ is an integral domain and let a, b, c be elements such that $ab = ac, a \neq 0$. Then $a(b - c) = 0$. Hence $b - c = 0$ and $b = c$. Similarly we can prove the right cancellation law. On the other hand, let $\mathfrak{A}$ be any ring for which the left cancellation law holds. Let $ab = 0$, $a \neq 0$. Then $ab = a0$ and $b = 0$. Hence $\mathfrak{A}$ is an integral domain.

A ring is called a *division ring* (*quasi-field, skew field, sfield*) if it contains more than one element and the set $\mathfrak{A}^*$ of non-zero elements forms a subgroup of the multiplicative semi-group. Thus if $\mathfrak{A}$ is a division ring, $\mathfrak{A}^*$ contains an identity element 1. Since $10 = 0 = 01$, 1 is an identity for the whole ring. Hence a division ring possesses an identity. Also if $a \neq 0$, then there exists an element a^{-1} in $\mathfrak{A}$ such that $aa^{-1} = 1 = a^{-1}a$. Examples (2), (3), (5), (6) and (9) are division rings in which multiplication is commutative. Division rings that have this property are called *fields*. We shall give an example of a non-commutative division ring in § 5.

It is clear from the definitions that any division ring is an integral domain. On the other hand, the converse does not hold since the ring I of integers is an integral domain but not a division ring. If $a \neq 0$ in a division ring $\mathfrak{A}$, then the equation $ax = b$ has the solution $x = a^{-1}b$ in $\mathfrak{A}$. By the restricted cancellation law this is the only solution of the equation. Similarly $ya = b$ has one and only one solution, namely, $y = ba^{-1}$.

Now let $\mathfrak{A}$ be any ring with an identity $1 \neq 0$. Our discussion of semi-groups shows that the totality $\mathfrak{U}$ of units of the multiplicative semi-group of $\mathfrak{A}$ is a subgroup of this semi-group. This means that the product of units is a unit, 1 is a unit and the inverse of a unit is a unit. We shall call $\mathfrak{U}$ the *group of units* of the ring $\mathfrak{A}$. For example, the group of units of I consists of the numbers 1 and -1. It is immediate that a ring $\mathfrak{A}$ is a division ring if and only if (1) $\mathfrak{A}$ contains an identity $\neq 0$ and (2) the group of units of $\mathfrak{A}$ is the set $\mathfrak{A}^*$ of non-zero elements.

EXERCISES

1. Prove that, if a is a unit in a ring with an identity, then so is $-a$. Show that $(-a)^{-1} = -a^{-1}$.

2. Show that the example given in ex. 2, p. 51, is a field.

3. Prove that any finite integral domain is a division ring.

4. Prove that, if an integral domain $\mathfrak{A}$ has an idempotent element $e \neq 0$ ($e^2 = e$), then e is an identity for $\mathfrak{A}$.

5. An element z of a ring is called *nilpotent* if $z^n = 0$. Show that the only nilpotent element of an integral domain is $z = 0$.

6. Show that, if a ring has only one left identity 1_l, then 1_l is an identity (two-sided).

7. Let u be an element of a ring with an identity that has a right inverse. Prove that the following conditions on u are equivalent: (1) u has more than one right inverse, (2) u is not a unit, (3) u is a left zero-divisor.

8. (Kaplansky.) Prove that, if an element of a ring with an identity has more than one right inverse, then it has infinitely many.

*3. Quasi-regularity. The circle composition.

***3. Quasi-regularity. The circle composition.** As we shall see, the groups of units of rings with identities give us interesting examples of groups. It is therefore noteworthy that the concept of the group of units has an analogue also for arbitrary rings that need not have identities. In order to obtain this, we assume first that $\mathfrak{A}$ has an identity. If a is an element of $\mathfrak{A}$ that has a right inverse b, then we may write $a = 1 - z$ and $b = 1 - w$ and obtain

$$1 = ab = (1 - z)(1 - w) = 1 - z - w + zw.$$

Hence the condition on z and w is that

$$z + w - zw = 0.$$

Since this condition does not involve the identity, we can use it for an arbitrary ring. Thus we say that the element z of $\mathfrak{A}$ is *right (left) quasi-regular* if there exists an element w in $\mathfrak{A}$ such that $z + w - zw = 0$ ($z + w - wz = 0$). The element w is called a *right (left) quasi-inverse* of z.

A still better insight into the concept of quasi-regularity is obtained by the following considerations. Let $\mathfrak{A}$ be an arbitrary ring and define a binary composition in $\mathfrak{A}$ by the formula

$$a \cdot b = a + b - ab.$$

We call this composition the *circle composition* in $\mathfrak{A}$. One verifies directly that it is associative; hence, $\mathfrak{A},\cdot$ is a semi-group. Also clearly $a \cdot 0 = a = 0 \cdot a$; hence, 0 acts as identity in $\mathfrak{A},\cdot$. It is now clear that the set of elements $\mathfrak{Q}$ that are *quasi-regular* ($=$ left and right quasi-regular) is just the set of units of $\mathfrak{A},\cdot$. Hence $\mathfrak{Q},\cdot$ is a group.

The group $\mathfrak{Q},\cdot$ is the analogue for an arbitrary ring of the group of units $\mathfrak{U}$ of a ring with an identity. In fact, if $\mathfrak{A}$ has an identity, then $\mathfrak{U}$ and $\mathfrak{Q}$ are isomorphic; for it is easy to see that the mapping $z \rightarrow 1 - z$ is an isomorphism of $\mathfrak{Q}$ onto $\mathfrak{U}$.

EXERCISES

1. Show that, if e is idempotent, then $e \cdot e = e$. Hence prove that, if e is right quasi-regular, then $e = 0$.

2. Show that any nilpotent element belongs to $\mathfrak{Q}$.

3. (Kaplansky.) Establish the following characterization of a division ring: A ring in which every element with one exception has a right quasi-inverse.

4. Matrix rings. Let $\mathfrak{R}$ be an arbitrary ring. We shall now define the ring $\mathfrak{R}_n$ of $n \times n$ matrices with elements in $\mathfrak{R}$. The elements of $\mathfrak{R}_n$ are arrays or *matrices*

$$(4) \qquad (a) = \begin{bmatrix} a_{11} & a_{12} & \cdots & a_{1n} \\ a_{21} & a_{22} & \cdots & a_{2n} \\ \cdot & \cdot & \cdots & \cdot \\ a_{n1} & a_{n2} & \cdots & a_{nn} \end{bmatrix}$$

of n rows and columns with *elements (coefficients, coordinates)* a_{ij} in the base ring $\mathfrak{R}$. The element a_{ij} in the intersection of the ith row and jth column of (a) will be referred to as the (i,j) *element* of (a). Two matrices (a) and (b) are regarded as equal if and only if $a_{ij} = b_{ij}$ for every i,j, and the set $\mathfrak{R}_n$ is the complete set of matrices with elements in $\mathfrak{R}$.

We define addition of matrices by the formula

$$\begin{bmatrix} a_{11} & a_{12} & \cdots & a_{1n} \\ a_{21} & a_{22} & \cdots & a_{2n} \\ \cdot & \cdot & \cdots & \cdot \\ a_{n1} & a_{n2} & \cdots & a_{nn} \end{bmatrix} + \begin{bmatrix} b_{11} & b_{12} & \cdots & b_{1n} \\ b_{21} & b_{22} & \cdots & b_{2n} \\ \cdot & \cdot & \cdots & \cdot \\ b_{n1} & b_{n2} & \cdots & b_{nn} \end{bmatrix}$$

$$= \begin{bmatrix} a_{11} + b_{11} & a_{12} + b_{12} & \cdots & a_{1n} + b_{1n} \\ a_{21} + b_{21} & a_{22} + b_{22} & \cdots & a_{2n} + b_{2n} \\ \cdot & \cdot & \cdots & \cdot \\ a_{n1} + b_{n1} & a_{n2} + b_{n2} & \cdots & a_{nn} + b_{nn} \end{bmatrix}.$$

Thus to obtain the sum we add the elements a_{ij} and b_{ij} in the same position. It is easy to verify that $\mathfrak{R}_n$ and this addition composition form a commutative group. The 0 matrix is the matrix all of whose elements are 0 and the negative of (a) has $-a_{ij}$ in the (i,j)-position, that is, in the intersection of the ith row and the jth column. Multiplication of matrices is defined by

$$\begin{bmatrix} a_{11} & a_{12} & \cdots & a_{1n} \\ a_{21} & a_{22} & \cdots & a_{2n} \\ \cdot & \cdot & \cdots & \cdot \\ a_{n1} & a_{n2} & \cdots & a_{nn} \end{bmatrix} \begin{bmatrix} b_{11} & b_{12} & \cdots & b_{1n} \\ b_{21} & b_{22} & \cdots & b_{2n} \\ \cdot & \cdot & \cdots & \cdot \\ b_{n1} & b_{n2} & \cdots & b_{nn} \end{bmatrix}$$

$$= \begin{bmatrix} \Sigma a_{1k} b_{k1} & \Sigma a_{1k} b_{k2} & \cdots & \Sigma a_{1k} b_{kn} \\ \Sigma a_{2k} b_{k1} & \Sigma a_{2k} b_{k2} & \cdots & \Sigma a_{2k} b_{kn} \\ \cdot & \cdot & \cdots & \cdot \\ \Sigma a_{nk} b_{k1} & \Sigma a_{nk} b_{k2} & \cdots & \Sigma a_{nk} b_{kn} \end{bmatrix}.$$

The product $(p) = (a)(b)$ therefore has the element

$$p_{ij} = a_{i1} b_{1j} + a_{i2} b_{2j} + \cdots + a_{in} b_{nj}$$

in the (i,j)-position. For example, in the ring I_3, I the ring of integers we have

$$\begin{bmatrix} 1 & -2 & 3 \\ 0 & 1 & -1 \\ 2 & 5 & -2 \end{bmatrix} \begin{bmatrix} 0 & 3 & 4 \\ 2 & 5 & 1 \\ -1 & -6 & 2 \end{bmatrix} = \begin{bmatrix} -7 & -25 & 8 \\ 3 & 11 & -1 \\ 12 & 43 & 9 \end{bmatrix}.$$

Multiplication of matrices is associative. Thus consider the product $(a)[(b)(c)]$. The multiplication rule shows that the element in the (i,j)-position of this matrix is $\sum_{k,l} a_{ik}(b_{kl}c_{lj})$. Similarly, the element in the (i,j)-position of $[(a)(b)](c)$ is $\sum_{k,l} (a_{ik}b_{kl})c_{lj}$. Because of the associative law of multiplication in $\mathfrak{R}$, these elements are equal. Hence $(a)[(b)(c)] = [(a)(b)](c)$. The distributive laws hold; for the (i,j)-element of $(a)[(b) + (c)]$ is $\sum_k a_{ik}(b_{kj} + c_{kj})$ and the (i,j) element of $(a)(b) + (a)(c)$ is $\sum_k a_{ik}b_{kj} + \sum_k a_{ik}c_{kj}$. These elements are equal by the distributive law in $\Re$. Similarly we can verify the other distributive law.

Hence $\Re_n$ is a ring. Even if $\Re$ is commutative, $\Re_n$ will not be commutative if $n > 1$ (cf. ex. 3 below). Also $\Re_n$ contains zero-divisors $\neq 0$ if $n > 1$.

EXERCISES

1. Calculate

$$\begin{bmatrix} 1 & -2 & 3 \\ -2 & 1 & 3 \\ 0 & 1 & -1 \end{bmatrix} \begin{bmatrix} 3 & -5 & 6 \\ 7 & 2 & 1 \\ -1 & 1 & 2 \end{bmatrix}.$$

2. Give examples to show that I_2 is not commutative and that it has zero-divisors $\neq 0$.

3. Prove that, if $\Re \neq 0$ and $n > 1$, then $\Re_n$ has zero-divisors $\neq 0$ and that, if $\Re$ contains elements a, b such that $ab \neq 0$, then $\Re_n$, $n > 1$, is not commutative.

If $\Re$ has an identity 1, then it is clear that the element

$$(5) \qquad 1 = \begin{bmatrix} 1 & & & 0 \\ & 1 & & \\ & & \ddots & \\ 0 & & & 1 \end{bmatrix}$$

is the identity in the ring $\Re_n$. We assume now that $\Re$ is commutative and we propose to determine the multiplicative group of units of $\Re_n$. For this purpose we make use of the determinant of a matrix. We assume that the reader is acquainted with the elementary theory of determinants of any order. The usual treatments in textbooks on elementary algebra or geometry are valid for determinants of matrices with elements in any commutative ring.

We recall here the definition of the *determinant* of a matrix. If (a) is as in (4) its determinant det (a) is

$$(6) \qquad \sum_P \pm a_{1i_1}a_{2i_2} \cdots a_{ni_n}$$

where the summation is taken over all permutations $(i_1, i_2, \cdots, i_n)$ of $(1, 2, \cdots, n)$ and the sign $+$ or $-$ is taken according as the permutation is even or odd. The *cofactor* of the element a_{ij} in (4) is $(-1)^{i+j}$ times the determinant of order $n - 1$ that is obtained by striking out the ith row and jth column of (a). It is well known that the sum of the products of the elements of any row (column) by their cofactors has the value det (a). Thus if A_{ij} is the cofactor of a_{ij}, then

(7)
$$a_{i1}A_{i1} + a_{i2}A_{i2} + \cdots + a_{in}A_{in} = \det (a)$$
$$a_{1i}A_{1i} + a_{2i}A_{2i} + \cdots + a_{ni}A_{ni} = \det (a).$$

Also it is known that the sum of the products of the elements of any row (column) by the cofactors of the elements of another row (column) is 0:

(8)
$$a_{i1}A_{j1} + a_{i2}A_{j2} + \cdots + a_{in}A_{jn} = 0, \quad i \neq j$$
$$a_{1i}A_{1j} + a_{2i}A_{2j} + \cdots + a_{ni}A_{nj} = 0, \quad i \neq j.$$

These relations lead us to define the *adjoint* of the matrix (a) to be the matrix whose (i, j) element $\alpha_{ij} = A_{ji}$. Using this definition it is immediate that the rules (7) and (8) are equivalent to the matrix equations

$$(9) \quad (a)\mathrm{adj}(a) = \begin{bmatrix} \det (a) & & & & 0 \\ & \det (a) & & & \\ & & \ddots & & \\ & & & & \\ 0 & & & & \det (a) \end{bmatrix} = [\mathrm{adj}(a)](a).$$

It follows that if $\Delta = \det (a)$ is a unit in $\mathfrak{R}$, then the matrix (b), $b_{ij} = \alpha_{ij}\Delta^{-1}$ satisfies

$$(10) \qquad\qquad (a)(b) = 1 = (b)(a).$$

We have therefore proved the sufficiency part of the following

Theorem 1. *If $\mathfrak{R}$ is a commutative ring with an identity, a matrix $(a) \varepsilon \mathfrak{R}_n$ is a unit if and only if its determinant is a unit in $\mathfrak{R}$.*

To prove the necessity we require the fundamental multiplication rule

$$(11) \qquad \det (a)(b) = \det (a) \det (b).$$

If $(a)(b) = 1$, then this gives $\det (a) \det (b) = 1$. Hence $\det (a)$ is a unit.

A noteworthy special case of this theorem is the

Corollary. *If $\mathfrak{R} = \mathfrak{F}$ is a field, a matrix $(a) \, \varepsilon \, \mathfrak{F}_n$ is a unit if and only if its determinant is different from zero.*

EXERCISES

1. Find the adjoint of the matrix

$$\begin{bmatrix} -1 & 2 & 4 \\ 3 & 2 & 0 \\ 5 & -1 & 2 \end{bmatrix}.$$

2. Show that the matrix

$$\begin{bmatrix} 1 & 4 & 1 \\ 0 & 1 & -1 \\ -3 & -6 & -8 \end{bmatrix}$$

is a unit in I_3, I the ring of integers. Find the inverse.

3. Prove that, if $\mathfrak{R}$ is a commutative ring with an identity, then $(a)(b) = 1$ for (a), (b) in $\mathfrak{R}_n$ implies that $(b)(a) = 1$.

5. Quaternions. We consider the set Q of matrices in C_2, C the field of complex numbers, that have the form

$$(12) \quad \begin{bmatrix} a & b \\ -b & \bar{a} \end{bmatrix} \equiv \begin{bmatrix} \alpha_0 + \alpha_1 \sqrt{-1} & \alpha_2 + \alpha_3 \sqrt{-1} \\ -\alpha_2 + \alpha_3 \sqrt{-1} & \alpha_0 - \alpha_1 \sqrt{-1} \end{bmatrix}, \quad \alpha_i \text{ real.}$$

We wish to show that Q determines a subgroup of the additive group of C_2 and that Q is closed under multiplication. The first of these assertions is easy to verify. Since

$$\begin{bmatrix} a & b \\ -\bar{b} & \bar{a} \end{bmatrix} \begin{bmatrix} c & d \\ -\bar{d} & \bar{c} \end{bmatrix} = \begin{bmatrix} ac - b\bar{d} & ad + b\bar{c} \\ -\bar{b}c - \bar{a}\bar{d} & \bar{a}\bar{c} - \bar{b}d \end{bmatrix},$$

the product has the form

$$\begin{bmatrix} u & v \\ -\bar{v} & \bar{u} \end{bmatrix}$$

where $u = ac - b\bar{d}$, $v = ad + b\bar{c}$. Hence it belongs to Q. Since the associative laws, the commutative law of addition and the distributive laws carry over from C_2 to the subset Q, it is clear that the system $Q, +, \cdot$ is a ring. Thus $Q, +, \cdot$ is an instance of a subring of the ring $C_2, +, \cdot$ in the sense of the following

Definition 2. *If $\mathfrak{B}$ is a subset of a ring $\mathfrak{A}$ that is closed under the compositions of the ring and $\mathfrak{B}, +, \cdot$ (induced compositions) is a ring, then $\mathfrak{B}, +, \cdot$ is called a subring of $\mathfrak{A}, +, \cdot$.*

As in the special case considered here it is clear that a subset $\mathfrak{B}$ determines a subring if $\mathfrak{B}, +$ is a group and $\mathfrak{B}$ is closed under multiplication. Also we recall that the first of these conditions is satisfied if either (1) $\mathfrak{B}$ is closed under $+$, contains 0 and the negative of any element in $\mathfrak{B}$ or (2) $\mathfrak{B}$ is closed under subtraction.

We shall now show that Q is a division ring. We note first that

$$\det \begin{bmatrix} \alpha_0 + \alpha_1 \sqrt{-1} & \alpha_2 + \alpha_3 \sqrt{-1} \\ -\alpha_2 + \alpha_3 \sqrt{-1} & \alpha_0 - \alpha_1 \sqrt{-1} \end{bmatrix}$$
$$= \alpha_0{}^2 + \alpha_1{}^2 + \alpha_2{}^2 + \alpha_3{}^2 \neq 0$$

if the matrix is $\neq 0$. Hence this matrix has an inverse. We determine it by the method of the preceding section, and we find that it is the matrix

$$\begin{bmatrix} (\alpha_0 - \alpha_1 \sqrt{-1}\,)\Delta^{-1} & -(\alpha_2 + \alpha_3 \sqrt{-1}\,)\Delta^{-1} \\ (\alpha_2 - \alpha_3 \sqrt{-1}\,)\Delta^{-1} & (\alpha_0 + \alpha_1 \sqrt{-1}\,)\Delta^{-1} \end{bmatrix}$$

where $\Delta = \alpha_0{}^2 + \alpha_1{}^2 + \alpha_2{}^2 + \alpha_3{}^2$. Thus the inverse is in Q. We have therefore shown that any non-zero element of Q has an inverse in Q. Hence Q *is a division ring*. We call Q *the ring of* (Hamilton's) *quaternions* and we call the elements of Q *quaternions*.

The ring Q contains the subring R' of matrices of the form

$$(13) \qquad \alpha' = \begin{bmatrix} \alpha & 0 \\ 0 & \alpha \end{bmatrix}.$$

It is easy to see that these matrices commute with every matrix in C_2 and hence with every quaternion. Also we note that the matrices

$$(14)\quad i = \begin{bmatrix} \sqrt{-1} & 0 \\ 0 & -\sqrt{-1} \end{bmatrix}, \; j = \begin{bmatrix} 0 & 1 \\ -1 & 0 \end{bmatrix}, \; k = \begin{bmatrix} 0 & \sqrt{-1} \\ \sqrt{-1} & 0 \end{bmatrix}$$

are quaternions. One verifies that

$$\begin{bmatrix} \alpha_0 + \alpha_1\sqrt{-1} & \alpha_2 + \alpha_3\sqrt{-1} \\ -\alpha_2 + \alpha_3\sqrt{-1} & \alpha_0 - \alpha_1\sqrt{-1} \end{bmatrix} = \alpha_0' + \alpha_1'i + \alpha_2'j + \alpha_3'k.$$

Hence if $\alpha_0' + \alpha_1'i + \alpha_2'j + \alpha_3'k = \beta_0' + \beta_1'i + \beta_2'j + \beta_3'k$, then

$$\begin{bmatrix} \alpha_0 + \alpha_1\sqrt{-1} & \alpha_2 + \alpha_3\sqrt{-1} \\ -\alpha_2 + \alpha_3\sqrt{-1} & \alpha_0 - \alpha_1\sqrt{-1} \end{bmatrix}$$

$$= \begin{bmatrix} \beta_0 + \beta_1\sqrt{-1} & \beta_2 + \beta_3\sqrt{-1} \\ -\beta_2 + \beta_3\sqrt{-1} & \beta_0 - \beta_1\sqrt{-1} \end{bmatrix}$$

and $\alpha_i = \beta_i$ and $\alpha_i' = \beta_i'$. This shows that the representation of a quaternion in the form $\alpha_0' + \alpha_1'i + \alpha_2'j + \alpha_3'k$ is unique. Since

$$(15)\qquad (\alpha + \beta)' = \alpha' + \beta', \quad (\alpha\beta)' = \alpha'\beta',$$

the product

$$(\alpha_0' + \alpha_1'i + \alpha_2'j + \alpha_3'k)(\beta_0' + \beta_1'i + \beta_2'j + \beta_3'k)$$

is determined by the addition and the multiplication in $\mathfrak{R}$ and by the multiplication table

$$i^2 = j^2 = k^2 = -1',$$

$$(16)$$

$$ij = -ji = k, \quad jk = -kj = i, \quad ki = -ik = j.$$

Incidentally these show that Q is not commutative. We remark finally that we can simplify our notation somewhat by replacing α' by α and more generally $\alpha_0' + \alpha_1'i + \alpha_2'j + \alpha_3'k$ by $\alpha_0 + \alpha_1 i + \alpha_2 j + \alpha_3 k$. We adopt this change in the following exercises.

EXERCISES

1. Calculate $(-1 + 2i - 3j + k)(2 - i + 3j - 2k)$.

2. Define the *trace* $T(a)$ of $a = \alpha_0 + \alpha_1 i + \alpha_2 j + \alpha_3 k$ to be the number $2\alpha_0$ and the *norm* $N(a) = \Delta = \alpha_0{}^2 + \alpha_1{}^2 + \alpha_2{}^2 + \alpha_3{}^2$. Verify that a satisfies the quadratic equation $x^2 - T(a)x + N(a) = 0$.

3. Prove that $N(ab) = N(a)N(b)$.

4. Show that the set Q_0 of quaternions $\alpha_0 + \alpha_1 i + \alpha_2 j + \alpha_3 k$ with rational coefficients α_i is a *division subring* of Q, that is, a subring that is a division ring.

5. Verify that the set J of quaternions $\alpha_0 + \alpha_1 i + \alpha_2 j + \alpha_3 k$ in which the α_i are either all integers or all halves of odd integers is a subring of Q. Is J a division subring?

6. Subrings generated by a set of elements. Center. It is clear from the definition of a subring that, if a number of subsets of a ring determine subrings, then their intersection has this property too. We express this somewhat more briefly by saying that the intersection of any collection of subrings of a ring is a subring. If S is any subset of the ring $\mathfrak{A}$, the intersection of the subrings containing S is called the *subring generated by S*. We denote this ring by $[[S]]$. Evidently $[[S]]$ is characterized by the following properties: (1) $[[S]]$ is a subring; (2) $[[S]] \supseteq S$; (3) if $\mathfrak{B}$ is any subring containing S, then $\mathfrak{B} \supseteq [[S]]$. It is easy to indicate the form of the elements of $[[S]]$, namely, they are the elements $\Sigma \pm s_1 s_2 \cdots s_r$, that is, the sums of finite products of elements s_i in S and negatives of such products; for the collection of such sums is a subring and it is clear that it has the properties (2) and (3) of $[[S]]$.

If S is a set of elements, the totality $C(S)$ of elements c that commute with every $s \, \varepsilon \, S$ is a subring. Evidently if $S_1 \supseteq S_2$, then $C(S_1) \subseteq C(S_2)$ and $C(C(S)) \supseteq S$. These two relations have the interesting consequence that

$$C(C(C(S))) = C(S);$$

for replacing S by $C(S)$ in $C(C(S)) \supseteq S$ gives $C(C(C(S))) \supseteq C(S)$. On the other hand, if we "operate" with C on both sides of this same relation we obtain $C(C(C(S))) \subseteq C(S)$.

If we refer to the form of the elements of $[[S]]$, we see that an element c that commutes with every element of S commutes also with every element of $[[S]]$. Hence $C(S) = C([[S]])$.

The subring $\mathfrak{C} = C(\mathfrak{A})$ is called the *center* of the ring. If $\mathfrak{A}$ contains an identity 1, evidently $1 \in \mathfrak{C}$.

EXERCISES

1. Determine the center of the ring of quaternions.
2. Let the α_i in

$$(\alpha) = \begin{bmatrix} \alpha_1 & & & 0 \\ & \alpha_2 & & \\ & & \ddots & \\ 0 & & & \alpha_n \end{bmatrix}$$

be distinct rational numbers. Show that $C(\alpha)$ in the matrix ring R_{0n}, R_0 the field of rational numbers, is the set of *diagonal matrices*, that is, the set of matrices that have the same form as (α).

3. Show that the center of R_{0n} is the set of *scalar* matrices

$$\begin{bmatrix} \alpha & & & 0 \\ & \alpha & & \\ & & \ddots & \\ 0 & & & \alpha \end{bmatrix}.$$

4. Find $C(S)$ in I_2 for S the set of matrices of the form $\begin{bmatrix} a & b \\ 0 & c \end{bmatrix}$.

7. Ideals, difference rings. Let $\mathfrak{B}$ be a subgroup of the additive group of $\mathfrak{A}$. Since addition is commutative, $\mathfrak{B}$ is an invariant subgroup and

$$(17) \qquad (a + \mathfrak{B}) + (c + \mathfrak{B}) = (a + c) + \mathfrak{B}$$

where addition is the addition defined for subsets. (We recall that $U + V$ is the totality of elements $u + v$, u in U and v in V.) The set $\bar{\mathfrak{A}} \equiv \mathfrak{A}/\mathfrak{B}$ of cosets is a commutative group relative to this composition. We now raise the following question: What is the condition on $\mathfrak{B}$ in order that $a \equiv a' \pmod{\mathfrak{B}}$ and $c \equiv c' \pmod{\mathfrak{B}}$ implies $ac \equiv a'c' \pmod{\mathfrak{B}}$ for all a, a', c, c'? If a and c are chosen, then $a' = a + b_1$ and $c' = c + b_2$ where b_1 and b_2 are in $\mathfrak{B}$. Also it is clear that any choice of b_1 and b_2 gives an $a' \equiv a \pmod{\mathfrak{B}}$ and a $c' \equiv c \pmod{\mathfrak{B}}$. Hence our requirement is equivalent to

$$(a + b_1)(c + b_2) = ac + ab_2 + b_1c + b_1b_2 \equiv ac \pmod{\mathfrak{B}}$$

for all a and c in $\mathfrak{A}$ and all b_1, b_2 in $\mathfrak{B}$. Thus

$$(18) \qquad\qquad ab_2 + b_1c + b_1b_2 \; \varepsilon \; \mathfrak{B}$$

for all a and c in $\mathfrak{A}$ and all b_1, b_2 in $\mathfrak{B}$. Taking $b_1 = 0$ this gives

$$(L) \qquad\qquad ab \; \varepsilon \; \mathfrak{B} \quad \text{for all } a \text{ in } \mathfrak{A} \text{ and all } b \text{ in } \mathfrak{B}$$

and taking $b_2 = 0$ this gives

$$(R) \qquad\qquad ba \; \varepsilon \; \mathfrak{B} \quad \text{for all } a \text{ in } \mathfrak{A} \text{ and all } b \text{ in } \mathfrak{B}.$$

Conversely if (L) and (R) hold, then ab_2, b_1c and $b_1b_2 \; \varepsilon \; \mathfrak{B}$ provided that b_1 and b_2 are in $\mathfrak{B}$. Hence (18) holds. This leads us to the important definition

Definition 3. *A subset $\mathfrak{B}$ of a ring $\mathfrak{A}$ is called an* ideal *if $\mathfrak{B},+$ is a subgroup of the additive group of $\mathfrak{A}$ and $\mathfrak{B}$ has the closure properties* (L) *and* (R).

Since a subset $\mathfrak{B}$ determines a subgroup if and only if the difference of every pair of its elements is contained in the set, we see that $\mathfrak{B}$ is an ideal if and only if (1) b_1, b_2 in $\mathfrak{B}$ imply that $b_1 - b_2 \; \varepsilon \; \mathfrak{B}$, (2) b in $\mathfrak{B}$ implies that ab and $ba \; \varepsilon \; \mathfrak{B}$ for all a in $\mathfrak{A}$. Evidently an ideal is closed under multiplication. Hence an ideal determines a subring of $\mathfrak{A}$.

If $\mathfrak{B}$ is an ideal in $\mathfrak{A}$, then our discussion shows that, if $a \equiv a'$ (mod $\mathfrak{B}$) and $c \equiv c'$ (mod $\mathfrak{B}$), then $ac \equiv a'c'$ (mod $\mathfrak{B}$). In other words, the product of any element in the coset $a + \mathfrak{B}$ by any element in the coset $c + \mathfrak{B}$ is an element in the coset $ac + \mathfrak{B}$. We can therefore define a (single-valued) multiplication composition for cosets by the formula

$$(19) \qquad\qquad (a + \mathfrak{B})(c + \mathfrak{B}) = ac + \mathfrak{B}.$$

It should be noted that this multiplication does not coincide with the multiplication of sets defined in the multiplicative semi-group. However, since we shall have no occasion to use the latter, no confusion will result from the notation in (19). We assert now that $\mathfrak{A}/\mathfrak{B}$, the addition (17) and the multiplication (19) constitute a ring. Since the rules for addition are clear we need only verify the associative and distributive laws. This is done in

$$[(a + \mathfrak{B})(c + \mathfrak{B})](d + \mathfrak{B}) = (ac + \mathfrak{B})(d + \mathfrak{B}) = (ac)d + \mathfrak{B}$$

$$(a + \mathfrak{B})[(c + \mathfrak{B})(d + \mathfrak{B})] = (a + \mathfrak{B})(cd + \mathfrak{B}) = a(cd) + \mathfrak{B}$$

and

$$(a + \mathfrak{B})[(c + \mathfrak{B}) + (d + \mathfrak{B})] = (a + \mathfrak{B})(c + d + \mathfrak{B})$$
$$= a(c + d) + \mathfrak{B}$$

$$(a + \mathfrak{B})(c + \mathfrak{B}) + (a + \mathfrak{B})(d + \mathfrak{B}) = (ac + \mathfrak{B}) + (ad + \mathfrak{B})$$
$$= (ac + ad) + \mathfrak{B}$$

and a similar calculation for the other distributive law. We call $\mathfrak{A}/\mathfrak{B}$ with the composition that we have defined the *difference (quotient, residue class) ring of $\mathfrak{A}$ relative to the ideal $\mathfrak{B}$.*

Some of the elementary properties of a ring carry over to any difference ring. Thus if $\mathfrak{A}$ is commutative then $\mathfrak{A}/\mathfrak{B}$ is commutative. This is clear from the definition. Similarly if $\mathfrak{A}$ has an identity 1, then $\bar{1} = 1 + \mathfrak{B}$ is an identity in $\mathfrak{A}/\mathfrak{B}$. On the other hand, we shall see in the next section that $\mathfrak{A}$ can be an integral domain and have difference rings that are not integral domains.

EXERCISES

1. Prove that, if n is any integer, then the set $n\mathfrak{A}$ of elements of the form na is an ideal.

2. Prove that the set of elements $\mathfrak{N}$ such that $na = 0$ is an ideal in any ring $\mathfrak{A}$.

8. Ideals and difference rings for the ring of integers. If m is any integer, the set (m) * of multiples of m is an ideal in the ring I of integers; for we know that (m) is a subgroup of the additive group and it is clear that a multiple of a multiple of m is a multiple of m. Also since the sets (m) are the only subgroups of I these are also the only ideals in the ring I. Since $(m) = (-m)$, we need consider only the cases $m = 0$ and $m > 0$. If $m = 0$, $(m) = 0$; hence $I/(m) = I$. Assume now that $m > 0$. Then we know that $I/(m)$ has the m elements

$$0 = \bar{0} = (m), \quad \bar{1} = 1 + (m), \quad \cdots, \quad \overline{(m - 1)} = m - 1 + (m).$$

The element $\bar{1} = 1 + (m)$ is the identity of $I/(m)$.

* Our group notation for this set is $[m]$.

Suppose first that m is composite, say, $m = m_1 m_2$ where the m_i are > 1. Then m_i is not divisible by m and $\bar{m}_i \neq 0$. On the other hand $\bar{m}_1 \bar{m}_2 = \overline{m_1 m_2} = \bar{m} = 0$. This shows that $I/(m)$ is not an integral domain.

Assume next that $m = p$ is *irreducible* (or *prime*) in the sense that p cannot be written as a product of integers greater than 1. In this case we can prove that $I/(p)$ is a field. We know that $I/(p)$ has an identity. Next let $\bar{a} \neq 0$. Then a is not divisible by p. Hence if $d = (a,p)$, $d \neq p$. Since p is prime, this leaves only the alternative $d = 1$. Hence there exist integers b and q such that $ab + pq = 1$. It follows that $\bar{a}\bar{b} = \overline{ab} = \bar{1}$. Hence $\bar{a}$ has the inverse $\bar{b}$ in $I/(p)$. Our result gives us the interesting conclusion that for any prime p there exists a field containing p elements.

We now drop the hypothesis that m is a prime, and we wish to determine the units in $I/(m)$. Let M denote the set of units and let $\bar{a} \, \varepsilon \, M$. Then there exists a $\bar{b}$ such that $\bar{a}\bar{b} = \bar{1}$. Hence $ab = 1 + mq$ and $ab - mq = 1$. This implies that $(a,m) = 1$. Conversely, if $(a,m) = 1$, then there exist b,q such that $ab - mq = 1$. Then $\bar{a}\bar{b} = 1$. This shows that in the list $\bar{0}, \bar{1}, \bar{2}, \cdots,$ $\overline{(m-1)}$ the units are the cosets $\bar{a}$ with $(a,m) = 1$ and it proves the following

Theorem 2. *The order of the group M of units of $I/(m)$ is the number of positive integers that are less than m and are relatively prime to m $((a,m) = 1)$.*

This number is denoted as $\phi(m)$ and the function of m thus determined is called Euler ϕ-*function* (*totient*).

We know that, if $\mathfrak{G}$ is a finite group of order n, then $a^n = 1$ for every $a \, \varepsilon \, \mathfrak{G}$. Applying this to M we see that, if $(a,m) = 1$, then $(\bar{a})^{\phi(m)} = \bar{1}$. The latter equation is equivalent to $a^{\phi(m)} \equiv 1 \pmod{m}$. Hence we have proved the following

Theorem 3 (Euler-Fermat). *If a is an integer prime to the positive integer m, then $a^{\phi(m)} \equiv 1 \pmod{m}$.*

If $m = p$, then $I/(p)$ is a field of p elements. The group of units in this case contains $p - 1$ elements. Hence we have the

Corollary. *If p is a prime and $a \not\equiv 0$ (mod p), then $a^{p-1} \equiv 1$ (mod p).*

This result can also be stated in a slightly different form, namely, that $a^p \equiv a$ (mod p). This holds for all a since it is trivial if a is divisible by p. On the other hand, if $a^p \equiv a$ (mod p) and $a \not\equiv 0$ (mod p), then $a^{p-1} \equiv 1$ (mod p). Hence the two statements are equivalent.

EXERCISE

1. Prove that, if D is a finite division ring containing q elements, then $a^q = a$ for every $a \varepsilon D$.

9. Homomorphism of rings

Definition 4. *A mapping η of a ring $\mathfrak{A}$ into a ring $\mathfrak{A}'$ is called a* homomorphism *if*

$$(a + b)\eta = a\eta + b\eta, \quad (ab)\eta = (a\eta)(b\eta).$$

Thus a homomorphism of a ring is a homomorphism of its additive group that "preserves" multiplication. If η is 1–1, it is called an *isomorphism* and two rings are said to be *isomorphic* ($\mathfrak{A} \cong \mathfrak{A}'$) if there exists an isomorphism of $\mathfrak{A}$ onto $\mathfrak{A}'$. As for groups it is immediate that the resultant of two homomorphisms is a homomorphism. Also if η is an isomorphism of $\mathfrak{A}$ onto $\mathfrak{A}'$, then the inverse mapping η^{-1} is an isomorphism of $\mathfrak{A}'$ onto $\mathfrak{A}$. It follows that the isomorphism relation is an equivalence relation in the class of rings. An isomorphism of a ring onto itself is called an *automorphism*. These concepts are illustrated in the following

EXERCISES

1. Show that the correspondence $\alpha + \beta\sqrt{-1} \to \begin{bmatrix} \alpha & \beta \\ -\beta & \alpha \end{bmatrix}$ is an isomorphism of the field C of complex numbers into R_2.

2. Show that the correspondence $a = \alpha + \beta\sqrt{-1} \to \bar{a} = \alpha - \beta\sqrt{-1}$ is an automorphism in C.

3. Show that the correspondence $\begin{bmatrix} \alpha & 0 \\ 0 & \beta \end{bmatrix} \to \alpha$ is a homomorphism of the ring of diagonal matrices into the ring coefficient $\mathfrak{R}$.

4. Show that the correspondence

$$\alpha_0 + \alpha_1 i + \alpha_2 j + \alpha_3 k \rightarrow \begin{bmatrix} \alpha_0 & \alpha_1 & \alpha_2 & \alpha_3 \\ -\alpha_1 & \alpha_0 & -\alpha_3 & \alpha_2 \\ -\alpha_2 & \alpha_3 & \alpha_0 & -\alpha_1 \\ -\alpha_3 & -\alpha_2 & \alpha_1 & \alpha_0 \end{bmatrix}$$

is an isomorphism of Q into R_4.

The theory of ring homomorphisms parallels that of group homomorphisms and in part is deducible from the latter theory. We begin our discussion by noting the following basic result.

Theorem 4. *If η is a homomorphism of $\mathfrak{A}$ into $\mathfrak{A}'$, the image set $\mathfrak{A}\eta$ is a subring of $\mathfrak{A}'$.*

Proof. Since η is a homomorphism of the additive group of $\mathfrak{A}$, $\mathfrak{A}\eta$ is a subgroup of the additive group of $\mathfrak{A}'$. Since $(a\eta)(b\eta) = (ab)\eta$, $\mathfrak{A}_\eta$ is closed under multiplication; hence it is a subring.

If the ring $\mathfrak{A}$ has an identity 1, then it is immediate that $1' = 1\eta$ is an identity for $\mathfrak{A}\eta$. Also if u is a unit with v as inverse, then $u' = u\eta$ is a unit in $\mathfrak{A}\eta$ with $v' = v\eta$ as its inverse. Of course, it may happen that $1\eta = 0$, but in this case $\mathfrak{A}\eta = 0$. In particular, if $\mathfrak{A}$ is a division ring, then either $\mathfrak{A}\eta = 0$ or $\mathfrak{A}\eta$ is also a division ring; for, if $\mathfrak{A}\eta \neq 0$, then this ring contains more than one element, and every non-zero element is a unit.

As for groups we call the inverse image $\eta^{-1}(0)$ the *kernel* of the homomorphism η. The homomorphism η is an isomorphism if and only if its kernel is 0.

Theorem 5. *The kernel of a homomorphism of a ring $\mathfrak{A}$ is an ideal in $\mathfrak{A}$.*

Proof. Let $\mathfrak{K} = \eta^{-1}(0)$. We know that $\mathfrak{K}$ is a subgroup of the additive group of $\mathfrak{A}$. Now let $b \,\varepsilon\, \mathfrak{K}$ and let a be arbitrary in $\mathfrak{A}$. Then $(ab)\eta = (a\eta)(b\eta) = (a\eta)0 = 0$. Hence $ab \,\varepsilon\, \mathfrak{K}$. Similarly $ba \,\varepsilon\, \mathfrak{K}$ and this completes the proof.

Next let $\mathfrak{B}$ be any ideal in the ring and let $\bar{\mathfrak{A}}$ denote the difference ring $\mathfrak{A}/\mathfrak{B}$. We know that the natural mapping ν is a homomorphism of the additive group of $\mathfrak{A}$ onto the additive group of $\bar{\mathfrak{A}}$. Moreover,

$$(a_1 a_2)\nu = a_1 a_2 + \mathfrak{B} = (a_1 + \mathfrak{B})(a_2 + \mathfrak{B}) = (a_1\nu)(a_2\nu).$$

Hence ν is a homomorphism of the ring $\mathfrak{A}$ onto the ring $\bar{\mathfrak{A}}$.

Now suppose that η is a homomorphism of the ring $\mathfrak{A}$ into the ring $\mathfrak{A}'$ with kernel $\mathfrak{K}$. Let $\mathfrak{B}$ be an ideal of $\mathfrak{A}$ contained in $\mathfrak{K}$. Then we know that the rule $a + \mathfrak{B} \to a\eta$ defines a homomorphism $\bar{\eta}$ of the additive group of $\bar{\mathfrak{A}} = \mathfrak{A}/\mathfrak{B}$ into the additive group of $\mathfrak{A}'$. Since

$$[(a_1 + \mathfrak{B})(a_2 + \mathfrak{B})]\bar{\eta} = (a_1a_2 + \mathfrak{B})\bar{\eta} = (a_1a_2)\eta = (a_1\eta)(a_2\eta)$$

$$= [(a_1 + \mathfrak{B})\bar{\eta}][(a_2 + \mathfrak{B})\bar{\eta}],$$

$\bar{\eta}$ is a ring homomorphism. Evidently $\eta = \nu\bar{\eta}$. We recall that $\bar{\eta}$ is 1–1 if and only if $\mathfrak{B} = \mathfrak{K}$. Thus if we take $\mathfrak{B} = \mathfrak{K}$, we obtain a factorization of η as $\nu\bar{\eta}$ where ν is the natural homomorphism of $\mathfrak{A}$ onto $\bar{\mathfrak{A}} = \mathfrak{A}/\mathfrak{K}$ and $\bar{\eta}$ is the induced isomorphism of $\bar{\mathfrak{A}}$ into $\mathfrak{A}'$. We state these results as the following important

Theorem 6. *Let η be a homomorphism of the ring $\mathfrak{A}$ into the ring $\mathfrak{A}'$ with kernel $\mathfrak{K}$ and let $\mathfrak{B}$ be an ideal of $\mathfrak{A}$ contained in $\mathfrak{K}$. Then the correspondence $\bar{\eta}$: $a + \mathfrak{B} \to a\eta$ is a homomorphism of $\bar{\mathfrak{A}} = \mathfrak{A}/\mathfrak{B}$ into $\mathfrak{A}'$ and $\eta = \nu\bar{\eta}$ where ν is the natural homomorphism of $\mathfrak{A}$ onto $\bar{\mathfrak{A}} = \mathfrak{A}/\mathfrak{B}$. The induced homomorphism $\bar{\eta}$ is an isomorphism if and only if $\mathfrak{B} = \mathfrak{K}$.*

If $\mathfrak{A}' = \mathfrak{A}\eta$ and $\mathfrak{B} = \mathfrak{K}$, then $\bar{\eta}$ is an isomorphism of $\bar{\mathfrak{A}}$ onto $\mathfrak{A}'$. This, together with an earlier result, gives the

Fundamental theorem of homomorphism of rings. *The difference ring $\mathfrak{A}/\mathfrak{B}$ of $\mathfrak{A}$ relative to any ideal $\mathfrak{B}$ is a homomorphic image of $\mathfrak{A}$. Conversely, any homomorphic image of $\mathfrak{A}$ is isomorphic to a difference ring, in fact, to the difference ring of $\mathfrak{A}$ relative to the kernel of the homomorphism.*

A ring $\mathfrak{A}$ is called *simple* if the only ideals in $\mathfrak{A}$ are $\mathfrak{A}$ and 0. (These are certainly ideals in any ring.) If $\mathfrak{A}$ has this property, then it is clear from the fundamental theorem that a homomorphic image of $\mathfrak{A}$ is either 0 or isomorphic to $\mathfrak{A}$.

As a second application of our results we determine next the structure of any ring $\mathfrak{A}$ that has an identity e and that is generated by e. We consider the ring of integers I and the mapping $n \to ne$ of I into $\mathfrak{A}$. Since

$$(n + m)e = ne + me$$

$$(nm)e = (nm)e^2 = (ne)(me),$$

our correspondence is a homomorphism. The image set Ie is a subring of $\mathfrak{A}$ including $1e = e$. Hence $Ie = \mathfrak{A}$ and $\mathfrak{A}$ is a homomorphic image of I. It follows that $\mathfrak{A} \cong I/(m)$ where $m \geq 0$. Thus either $\mathfrak{A}$ is infinite and isomorphic to the ring of integers or $\mathfrak{A}$ has a finite number m of elements and $\mathfrak{A}$ is isomorphic to the finite ring $I/(m)$.

EXERCISES

1. Let $m = rs \, \varepsilon \, I$. Show that $(r)/(m)$ is an ideal in $I/(m)$ and prove that

$$[I/(m)]/[(r)/(m)] \cong I/(r).$$

2. Determine the ideals and hence the homomorphic images of the subring of I_2 of matrices of the form $\begin{bmatrix} a & b \\ 0 & c \end{bmatrix}$.

3. Prove that, if $a \to \bar{a}$ is a homomorphism of $\mathfrak{R}$ into $\bar{\mathfrak{R}}$, then the mapping $(a_{ij}) \to (\bar{a}_{ij})$ is a homomorphism of $\mathfrak{R}_n$ into $\bar{\mathfrak{R}}_n$.

4. Let η be a homomorphism of a ring $\mathfrak{A}$ into itself. Show that the elements of $\mathfrak{A}$ that are *fixed* relative to η in the sense that $a\eta = a$ form a subring of $\mathfrak{A}$. If $\mathfrak{A}$ is a division ring and $\mathfrak{A}\eta \neq 0$, then the set of fixed elements constitutes a division subring.

5. Prove that the only homomorphisms of I into itself are the identity mapping and the mapping that sends every element into 0. Prove the same result for the field of rational numbers.

6. Let $\mathfrak{B}$ be a set and let η be a 1–1 mapping of $\mathfrak{B}$ onto a ring $\mathfrak{A}$. Prove that the compositions $a + b \equiv (a\eta + b\eta)\eta^{-1}$, $ab \equiv ((a\eta)(b\eta))\eta^{-1}$ turn $\mathfrak{B}$ into a ring isomorphic to $\mathfrak{A}$. Use this to prove that any ring is also a ring relative to the compositions $a \oplus b = a + b - 1, a \cdot b = a + b - ab$.

10. Anti-isomorphism. If a is the quaternion $\alpha_0 + \alpha_1 i + \alpha_2 j + \alpha_3 k$, we call the quaternion

$$\bar{a} = \alpha_0 - \alpha_1 i - \alpha_2 j - \alpha_3 k$$

the *conjugate* of a. If we refer to § 5 we can see that the inverse a^{-1} of $a \neq 0$ can be expressed in terms of the conjugate by means of the formula $a^{-1} = \bar{a}N(a)^{-1} = N(a)^{-1}\bar{a}$. We consider now the properties of the correspondence $a \to \bar{a}$. Evidently this mapping is 1–1 of Q onto itself. Also it is clear that

$$(20) \qquad \overline{a + b} = \bar{a} + \bar{b}$$

and we can verify that

$$\overline{ab} = (\alpha_0\beta_0 - \alpha_1\beta_1 - \alpha_2\beta_2 - \alpha_3\beta_3)$$
$$- (\alpha_0\beta_1 + \alpha_1\beta_0 + \alpha_2\beta_3 - \alpha_3\beta_2)i$$
$$- (\alpha_0\beta_2 + \alpha_2\beta_0 + \alpha_3\beta_1 - \alpha_1\beta_3)j$$
$$- (\alpha_0\beta_3 + \alpha_3\beta_0 + \alpha_1\beta_2 - \alpha_2\beta_1)k$$

and

$$\bar{b}\bar{a} = (\beta_0\alpha_0 - \beta_1\alpha_1 - \beta_2\alpha_2 - \beta_3\alpha_3)$$
$$+ (-\beta_0\alpha_1 - \beta_1\alpha_0 + \beta_2\alpha_3 - \beta_3\alpha_2)i$$
$$+ (-\beta_0\alpha_2 - \beta_2\alpha_0 + \beta_3\alpha_1 - \beta_1\alpha_3)j$$
$$+ (-\beta_0\alpha_3 - \beta_3\alpha_0 + \beta_1\alpha_2 - \beta_2\alpha_1)k.$$

Hence

(21) $$\overline{ab} = \bar{b}\bar{a}.$$

A mapping of a ring $\mathfrak{A}$ onto a ring $\bar{\mathfrak{A}}$ that is 1–1 and that satisfies (20) and (21) is called an *anti-isomorphism*. If $\bar{\mathfrak{A}}$ is commutative, then we can write $\bar{a}\bar{b}$ for $\bar{b}\bar{a}$ in (21) and we see that in this case $a \to \bar{a}$ is also an isomorphism of $\mathfrak{A}$ onto $\bar{\mathfrak{A}}$. Conversely any isomorphism between commutative rings can be regarded as an anti-isomorphism. In particular we see that the identity mapping is an anti-isomorphism of $\mathfrak{A}$ onto itself if $\mathfrak{A}$ is commutative. On the other hand, the quaternion example shows that there also exist non-commutative rings that have the symmetry property of being anti-isomorphic with themselves. We now give another important example of this type, namely, the matrix ring $\mathfrak{R}_n$, where $\mathfrak{R}$ is any commutative ring.

For this purpose we define the *transposed matrix* $(a)'$ of the matrix (a) to be the matrix that has a_{ji} in its (i,j) position. This means that $(a)'$ is obtained from (a) by reflecting the elements in the main diagonal. For example, if

$$(a) = \begin{bmatrix} 1 & 2 & -3 \\ 2 & -1 & 4 \\ 5 & -1 & 6 \end{bmatrix},$$

then

$$(a)' = \begin{bmatrix} 1 & 2 & 5 \\ 2 & -1 & -1 \\ -3 & 4 & 6 \end{bmatrix}.$$

In general if $(a) = (a_{ij})$, $(b) = (b_{ij})$, then $(a) + (b) = (a_{ij} + b_{ij})$, and $[(a) + (b)]'$ has the element $a_{ji} + b_{ji}$ in its (i,j)-position. Hence $[(a) + (b)]' = (a)' + (b)'$. Also the (i,j)-element of the product $(p) = (a)(b)$ is $p_{ij} = \sum_k a_{ik}b_{kj}$ so that the (i,j)-element of $(p)'$ is $p_{ji} = \Sigma a_{jk}b_{ki}$. On the other hand the (i,j)-element of $(b)'(a)'$ is $\Sigma b_{ki}a_{jk}$. Since we have assumed that $\Re$ is commutative, this shows that

$$[(a)(b)]' = (b)'(a)'.$$

Thus $(a) \to (a)'$, which is evidently 1–1, is an anti-isomorphism of $\Re_n$ onto itself.

We can construct for any given ring $\mathfrak{A},+,\cdot$ an anti-isomorphic ring. For this purpose we use the set $\mathfrak{A}$ and the given addition, but we introduce a new multiplication $\times$ defined by

$$a \times b = ba.$$

This gives a ring since

$$(a \times b) \times c = (ba) \times c = c(ba)$$

$$a \times (b \times c) = (b \times c)a = (cb)a$$

and

$$a \times (b + c) = (b + c)a = ba + ca = a \times b + a \times c$$

$$(b + c) \times a = a(b + c) = ab + ac = b \times a + c \times a.$$

Also it is immediate that the identity mapping is an anti-isomorphism of $\mathfrak{A},+,\cdot$ onto $\mathfrak{A},+,\times$.

EXERCISES

1. Show that the set of matrices of the form

$$\begin{bmatrix} a & b \\ 0 & 0 \end{bmatrix} \quad a,b \text{ in } I$$

is a subring of I_2 that has a left identity but no right identity. Hence prove that this ring is not anti-isomorphic to itself.

2. Define anti-isomorphism for semi-groups. Prove that any group is anti-isomorphic with itself.

3. An anti-isomorphism of a ring onto itself is usually called an *anti-automorphism*. Prove that the set of automorphisms and anti-automorphisms of a ring forms a transformation group. Show that the automorphisms form an invariant subgroup of index 1 or 2 in this group.

4. Show that, if $a \rightarrow \bar{a}$ is an anti-isomorphism of $\mathfrak{R}$ onto $\bar{\mathfrak{R}}$, then the mapping $(a) \rightarrow (\bar{a})'$, where the (i,j) element of $(\bar{a})'$ is $\bar{a}_{ji}$, is an anti-isomorphism of $\mathfrak{R}_n$ onto $\bar{\mathfrak{R}}_n$.

5. Define *anti-homomorphism*. State and prove the "fundamental theorem" for anti-homomorphisms.

6. (Hua) Let S be a mapping of a ring $\mathfrak{A}$ into a ring $\mathfrak{B}$ such that $(a + b)^S = a^S + b^S$ and for each pair a,b either $(ab)^S = a^S b^S$ or $(ab)^S = b^S a^S$. Prove that S is either a homomorphism or an anti-homomorphism.

11. Structure of the additive group of a ring. The characteristic of a ring.

If $\mathfrak{A},+$ is any commutative group, we can define a multiplication $ab = 0$ for all a,b and thus obtain a ring. It is clear that this composition is associative and distributive with respect to addition. A ring of this type is called a *zero-ring*. The existence of such rings shows that there is nothing that we can say in general about the structure of the additive group of a ring. However, as we proceed to show, simple restrictions imposed on the multiplicative semi-group of a ring will impose strong restrictions on the additive group.

For example, suppose that $\mathfrak{A}$ has an identity 1 and suppose 1 has finite order m in $\mathfrak{A},+$. Then if a is any element of $\mathfrak{A}$

$$ma = m(1a) = (m1)a = 0a = 0.$$

Hence every element has finite order a divisor of m.

If there exists a maximum m (>0) for the orders of the elements of $\mathfrak{A},+$, then the number m is called *the characteristic* of $\mathfrak{A}$. If no such maximum exists, we say that $\mathfrak{A}$ has *characteristic* 0 (or *infinity*).* Thus we see that, if $\mathfrak{A}$ has an identity 1, its characteristic is $m > 0$ or 0 according as 1 has order m or infinite order in $\mathfrak{A},+$.

We can generalize this result. Thus suppose that d is an element of $\mathfrak{A}$ that has finite order m and that d is not a left zero-divisor. If a is any element of $\mathfrak{A}$,

$$0 = (md)a = d(ma).$$

Hence $ma = 0$. Thus again the characteristic of $\mathfrak{A}$ is m. A similar result holds, of course, for elements that are not right zero-divisors.

---

* The terminology "characteristic infinity" is the more natural one from the present point of view. However, from another point of view (cf. pp. 103) "characteristic zero" is also natural. At any rate the latter seems to be the one that is most commonly used and we shall adopt it here.

In particular we see that, if $\mathfrak{A}$ is an integral domain, then either the characteristic is 0 or the characteristic is $m > 0$, and every non-zero element has order m. We shall now show that in the latter case m is a prime; for let $m = m_1 m_2$ where the $m_i > 1$. If $a \neq 0$,

$$ma^2 = m_1 m_2 a^2 = (m_1 a)(m_2 a).$$

Since $m_1 a \neq 0$ and $m_2 a \neq 0$, this is a contradiction. We have therefore proved the following

Theorem 7. *If $\mathfrak{A}$ is an integral domain of characteristic 0, then all of the non-zero elements of $\mathfrak{A}$ have infinite order. If $\mathfrak{A}$ has characteristic $m > 0$, then m is a prime and all of the non-zero elements of $\mathfrak{A}$ have order m.*

EXERCISE

1. Show that Theorem 7 holds for simple rings (instead of integral domains).

12. Algebra of subgroups of the additive group of a ring. One-sided ideals. We investigate in this section some important compositions that can be defined in the collection of subgroups of the additive group of a ring. Two of these, *intersection* and *the group generated by a collection of subgroups*, have been discussed for arbitrary groups. In the present situation the group that we start with is commutative; hence all subgroups are invariant. Hence, if A and B are subgroups, the subgroup $[A \cup B]$ generated by A and B coincides with the set $A + B$ of sums $a + b$, a in A, b in B. More generally, if $\{A_\alpha\}$ is a collection of subgroups of the additive group, then the group $[\cup A_\alpha]$ generated by the A_α is the set of finite sums

$$a_{\alpha_1} + a_{\alpha_2} + \cdots + a_{\alpha_k}, \quad a_{\alpha_i} \varepsilon A_{\alpha_i};$$

for it can be verified that the totality of these sums, which we denote now as ΣA_α, is a subgroup of the additive group. Also ΣA_α contains all the A_α and is contained in any subgroup that has this property. Hence ΣA_α has the properties that are characteristic for $[\cup A_\alpha]$.

We shall now introduce the third important composition on subgroups of the additive group. If A and B are subgroups, we define the *product* AB to be the subgroup generated by all of the products ab, a in A, b in B. It should be noted that this definition is different from the definition of multiplication for cosets. However, since the cosets $\neq B$ of a subgroup B are not subgroups, no real difficulty will result from the double use of the multiplication notation. We now note that AB coincides with the set P of finite sums

$$a_1b_1 + a_2b_2 + \cdots + a_kb_k$$

with a_i in A and b_i in B. It is clear that P contains all the products ab and that P is contained in any subgroup that contains all of these products. Also it is clear that P is closed under addition and that P contains 0. Finally, $-(a_1b_1 + \cdots + a_kb_k) = (-a)b_1 + \cdots + (-a_k)b_k \, \varepsilon \, P$. Hence P is a subgroup. These properties of P, of course, imply that $P = AB$.

We can easily establish the associative law $(AB)C = A(BC)$; for either of these subgroups is the totality of finite sums of the form $\Sigma a_i b_i c_i$, $a_i \, \varepsilon \, A$, $b_i \, \varepsilon \, B$, $c_i \, \varepsilon \, C$. Also we have the distributive laws $A(B + C) = AB + AC$ and $(B + C)A = BA + CA$. We prove the first of these by noting that $A(B + C)$ is the subgroup generated by all products $a(b + c)$, $a \, \varepsilon \, A$, $b \, \varepsilon \, B$, $c \, \varepsilon \, C$.

Since $a(b + c) = ab + ac \, \varepsilon \, AB + AC$, $A(B + C) \subseteq AB + AC$. On the other hand $ab = a(b + 0)$ is in $A(B + C)$. Hence $AB \subseteq A(B + C)$. Similarly $AC \subseteq A(B + C)$. But then $AB + AC \subseteq A(B + C)$. Hence $A(B + C) = AB + AC$. Evidently this same argument applies to the other distributive law.

The powers of a subgroup are defined inductively by $A^1 = A$, $A^k = (A^{k-1})A$. It is immediate that A^k is the set of finite sums of products of the form $a_1 a_2 \cdots a_k$ with the a_i in A. A subgroup A of the additive group determines a subring if and only if A is closed under multiplication. The condition for this can be expressed in terms of our multiplication as $A^2 \subseteq A$. The conditions that a subgroup $\mathfrak{B}$ be an ideal are that

(L) $$\mathfrak{A}\mathfrak{B} \subseteq \mathfrak{B}$$

(R) $$\mathfrak{B}\mathfrak{A} \subseteq \mathfrak{B}.$$

An important role is played in the theory of rings by subgroups that satisfy just one of the above conditions. If $\mathfrak{B}$ is a subgroup such that (L) holds, then $\mathfrak{B}$ is called a *left ideal* in $\mathfrak{A}$ and, if (R) holds, then $\mathfrak{B}$ is a *right ideal*.

Example. Let $\mathfrak{R}_n$ be the matrix ring defined by the ring $\mathfrak{R}$ and consider the subset $\mathfrak{B}$ of $\mathfrak{R}_n$ of matrices of the form

$$\begin{bmatrix} a_{11} & a_{12} & \cdots & a_{1k} & 0 & \cdots & 0 \\ a_{21} & a_{22} & \cdots & a_{2k} & 0 & \cdots & 0 \\ \cdot & \cdot & \cdots & \cdot & \cdot & \cdots & \cdot \\ a_{n1} & a_{n2} & \cdots & a_{nk} & 0 & \cdots & 0 \end{bmatrix}$$

where the a_{ij} are arbitrary. Then $\mathfrak{B}$ is a left ideal. Similarly the totality of matrices in which the last $n-k$ rows consist of 0's is a right ideal in $\mathfrak{R}_n$. It can be shown that neither of these is a (two-sided) ideal.

In any ring $\mathfrak{A}$ the totality $\mathfrak{A}b$ of left multiples xb, x in $\mathfrak{A}$, is a left ideal. If $\mathfrak{A}$ contains an identity, then $\mathfrak{A}b$ contains b and then $\mathfrak{A}b$ can be characterized as the smallest left ideal that contains b; for it is evident that $\mathfrak{A}b$ is contained in every left ideal that contains b. If $\mathfrak{A}$ does not have an identity, it is necessary to take the set of elements of the form $nb + xb$, n an integer, x arbitrary in $\mathfrak{A}$, to obtain the smallest left ideal containing b. In any case we shall call the smallest left ideal containing an element b a *principal left ideal*. We denote this ideal as $(b)_l$ so that $(b)_l = \mathfrak{A}b$ if $\mathfrak{A}$ has an identity and $(b)_l$ is the set $\{nb + xb\}$ for arbitrary $\mathfrak{A}$. In a similar manner we define the right ideal $b\mathfrak{A}$ of right multiples of b and the principal right ideal $(b)_r$. We always have $(b)_r \supseteq b\mathfrak{A}$ and $(b)_r = b\mathfrak{A}$ if $\mathfrak{A}$ has an identity.

The concept of a one-sided ideal can be used to give a new characterization of division rings:

Theorem 8. *A ring $\mathfrak{A}$ with an identity $1 \neq 0$ is a division ring if and only if it has no proper left (right)ideals.*

Proof. Suppose first that $\mathfrak{A}$ is a division ring. Then, if $\mathfrak{B}$ is a left ideal in $\mathfrak{A} \neq 0$, $\mathfrak{B}$ contains an element $b \neq 0$. Then $1 = b^{-1}b \, \varepsilon \, \mathfrak{B}$ and every $x = x1$ is in $\mathfrak{B}$. Hence $\mathfrak{B} = \mathfrak{A}$. Thus if $\mathfrak{B}$ is any left ideal, either $\mathfrak{B} = 0$ or $\mathfrak{B} = \mathfrak{A}$. Conversely let $\mathfrak{A}$ be a ring with an identity $1 \neq 0$ that has no proper left ideals. If b is an element $\neq 0$ in $\mathfrak{A}$, $\mathfrak{A}b$ contains $1b \neq 0$. Hence $\mathfrak{A}b = \mathfrak{A}$. This implies that there is a $c \, (\neq 0)$ such that $cb = 1$. Hence every element $\neq 0$ has a left inverse $\neq 0$ and this implies that the non-zero elements of $\mathfrak{A}$ form a group under multiplication (cf. ex. 2, p. 24). Hence $\mathfrak{A}$ is a division ring.

Of course, this result implies that any division ring is simple. It follows that the only homomorphic images of a division ring are 0 and the ring itself.

It can be verified that the compositions of intersection, sum and product applied to left (right) ideals give left (right) ideals. Other results of this type can be established. For example, the product $\mathfrak{B}\mathfrak{C}$ is a left ideal if $\mathfrak{B}$ is any left ideal and $\mathfrak{C}$ is a subgroup. Also $\mathfrak{B}\mathfrak{C}$ is a (two-sided) ideal if $\mathfrak{B}$ is a left ideal and $\mathfrak{C}$ is a right ideal.

EXERCISES

1. Prove that a ring $\mathfrak{A}$ which possesses no proper left ideals is either a division ring or a zero ring.

2. If $\mathfrak{A}$ is any ring, $\mathfrak{A}^2$, $\mathfrak{A}^3$, $\cdots$ are ideals. What are these ideals for the sub-ring of I_3 consisting of the matrices of the form

$$\begin{bmatrix} 0 & a & b \\ 0 & 0 & c \\ 0 & 0 & 0 \end{bmatrix} ?$$

13. The ring of endomorphisms of a commutative group. Let $\mathfrak{G}$ be an arbitrary commutative group. We use the additive notation in $\mathfrak{G}$: $+$ for the composition, 0 for the identity, $-a$ for the inverse and ma for the power or multiple of a. We consider now the set $\mathfrak{E}$ of endomorphisms of $\mathfrak{G}$. These are the mappings η of $\mathfrak{G}$ into itself such that

$$(22) \qquad\qquad (a + b)\eta = a\eta + b\eta.$$

We know that, if η, $\rho \, \varepsilon \, \mathfrak{E}$, then $\eta\rho \, \varepsilon \, \mathfrak{E}$ and the associative law holds for the resultant composition. We know also that the identity mapping belongs to $\mathfrak{E}$. These results hold even if $\mathfrak{G}$ is not commutative. However, a great deal more can be proved in the commutative case, namely, we can show that the set $\mathfrak{E}$ can be used to define a ring.

We introduce an addition composition in $\mathfrak{E}$ by defining $\eta + \rho$ by

$$(23) \qquad\qquad a(\eta + \rho) = a\eta + a\rho.$$

This mapping is an endomorphism since

$$(a + b)(\eta + \rho) = (a + b)\eta + (a + b)\rho$$
$$= a\eta + b\eta + a\rho + b\rho$$
$$= a\eta + a\rho + b\eta + b\rho$$
$$= a(\eta + \rho) + b(\eta + \rho).$$

It is easy to verify that $\mathfrak{E},+$ constitute a commutative group. We have $a(\eta + (\rho + \lambda)) = a\eta + a(\rho + \lambda) = a\eta + a\rho + a\lambda$ and $a((\eta + \rho) + \lambda) = a(\eta + \rho) + a\lambda = a\eta + a\rho + a\lambda$; hence $\eta + (\rho + \lambda) = (\eta + \rho) + \lambda$. Similarly $\eta + \rho = \rho + \eta$. We now define the 0 mapping to be the one which sends every a into 0. It is clear that this is an endomorphism and that $\eta + 0 = \eta$ for all η. Finally, if $\eta \, \varepsilon \, \mathfrak{E}$, we define $-\eta$ to be the mapping $a \to -(a\eta)$. This mapping may be regarded as the resultant of $a \to a\eta$ and the automorphism $a \to -a$. Hence $-\eta \, \varepsilon \, \mathfrak{E}$. Evidently $\eta + (-\eta) = 0$.

We shall now show that $\mathfrak{E},+,\cdot$ is a ring where the product $\cdot$ is the resultant. Since we know that $\mathfrak{E},+$ is a commutative group and since we know that $\cdot$ is associative, we have to prove only the distributive laws. Now we have

$$a(\eta(\rho + \lambda)) = (a\eta)(\rho + \lambda) = (a\eta)\rho + (a\eta)\lambda = a(\eta\rho) + a(\eta\lambda)$$
$$= a(\eta\rho + \eta\lambda),$$

so that $\eta(\rho + \lambda) = \eta\rho + \eta\lambda$ and

$$a((\rho + \lambda)\eta) = (a(\rho + \lambda))\eta = (a\rho + a\lambda)\eta = (a\rho)\eta + (a\lambda)\eta$$
$$= a(\rho\eta) + a(\lambda\eta) = a(\rho\eta + \lambda\eta).$$

Hence $(\rho + \lambda)\eta = \rho\eta + \lambda\eta$. This completes the proof of the following fundamental

Theorem 9. *Let $\mathfrak{G}$ be an arbitrary commutative group (written additively) and let $\mathfrak{E}$ be the totality of endomorphisms of $\mathfrak{G}$. Then $\mathfrak{E}$ is closed relative to the addition composition defined by $a(\eta + \rho) = a\eta + a\rho$ and relative to the resultant composition $\cdot$, and the system $\mathfrak{E},+,\cdot$ is a ring.*

We call $\mathfrak{E}$ *the ring of endomorphisms of* $\mathfrak{G}$. More generally we shall be interested in considering subrings of rings $\mathfrak{E}$. Such a subring will be called *a ring of endomorphisms* and we shall see in the next section that these rings play the same role in ring theory that transformation groups play in group theory. Before we discuss this, however, we consider some examples.

Examples. (1) $\mathfrak{G}$ an infinite cyclic group. Thus we can take $\mathfrak{G}$ to be the additive group I_+ of integers. If $\eta \, \epsilon \, \mathfrak{E}$ and $1\eta = u$ in I_+, then $n\eta = nu$ since η is an endomorphism. Now this remark shows that η is completely determined by its effect on the generator 1 of I_+. We shall therefore associate the integer u (effect of η on 1) with the endomorphism η. Suppose now that ρ is a second endomorphism and that $1\rho = v$. Then we associate v with ρ. Also $1(\eta + \rho) = 1\eta + 1\rho = u + v$ and $(1\eta)\rho = u\rho = uv$. Hence in our correspondence, $\eta + \rho \rightarrow u + v$ and $\eta\rho \rightarrow uv$. Also our correspondence is 1-1; for, if $u = v$ then $1\eta = 1\rho$ and since an endomorphism is determined by its effect on 1, $\eta = \rho$. Thus we have an isomorphism of $\mathfrak{E}$ into the *ring* of integers I. We remark finally that our isomorphism is one onto I. Thus if u is any integer, then the mapping $n \rightarrow nu$ is an endomorphism, since

$$(n + m)u = nu + mu$$

is a basic property of multiples. Clearly this endomorphism sends 1 into u. Thus we have proved that $\mathfrak{E}$ is isomorphic to I.

(2) As a generalization of (1) we consider next the group $\mathfrak{G}$ of all integral vectors $(m_1, m_2, \cdots, m_n)$, m_i in I. The composition here is vector addition. Hence if we introduce the vectors

$$\overset{i}{}$$
$$(24) \qquad e_i = (0, \cdots, 0, 1, 0, \cdots, 0), \quad i = 1, 2, \cdots, n,$$

then we can write

$$(25) \qquad (m_1, m_2, \cdots, m_n) = m_1 e_1 + m_2 e_2 + \cdots + m_n e_n.$$

Thus any integral vector is in the group generated by the e_i. Also it is clear that a vector can be written in only one way as $\Sigma m_i e_i$; for if $\Sigma m_i e_i = \Sigma m_i' e_i$, then by (25)

$$(m_1, m_2, \cdots, m_n) = (m_1', m_2', \cdots, m_n')$$

and $m_i = m_i'$ for all i.

Now let η be an endomorphism in $\mathfrak{G}$. We are going to show that η is completely determined by its effect on the e_i; for if the images $e_i\eta = f_i$ are known, then the image

$$(\Sigma m_i e_i)\eta = \Sigma(m_i e_i)\eta = \Sigma m_i(e_i\eta) = \Sigma m_i f_i$$

is known. It follows that, if η and ρ are two endomorphisms and $e_i\eta = e_i\rho$ for $i = 1, 2, \cdots, n$, then $a\eta = a\rho$ for all a. Hence $\eta = \rho$.

Suppose next *t*hat

$$(26) \qquad f_i = e_i\eta = a_{i1}e_1 + a_{i2}e_2 + \cdots + a_{in}e_n$$

where the a_{ij} are integers. It is clear that these integers are uniquely determined by η. Hence the matrix

$$(a_{ij}) = \begin{bmatrix} a_{11} & a_{12} & \cdots & a_{1n} \\ a_{21} & a_{22} & \cdots & a_{2n} \\ \cdot & \cdot & \cdots & \cdot \\ a_{n1} & a_{n2} & \cdots & a_{nn} \end{bmatrix}$$

is determined by η. We shall call this matrix *the matrix* of η, and we shall investigate the correspondence $\eta \to (a_{ij})$ of $\mathfrak{E}$ into the ring I_n of $n \times n$ matrices with elements in I.

We note first that our correspondence is 1–1; for if $\eta \to (a_{ij})$ and $\rho \to (a_{ij})$, then $e_i\eta = e_i\rho$ and hence $\eta = \rho$. Next let ρ be any endomorphism and let $\rho \to (b_{ij})$. Then $e_i\rho = \sum_j b_{ij}e_j$. Hence

$$e_i(\eta + \rho) = e_i\eta + e_i\rho = \sum_j a_{ij}e_j + \sum_j b_{ij}e_j$$

$$= \Sigma(a_{ij} + b_{ij})e_j.$$

Thus $\eta + \rho \to (a_{ij}) + (b_{ij})$. Finally

$$e_i(\eta\rho) = (e_i\eta)\rho = \left(\sum_j a_{ij}e_j\right)\rho = \sum_j (a_{ij}e_j)\rho = \sum_j a_{ij}(e_j\rho) = \sum_{j,k} a_{ij}b_{jk}e_k = \sum_k c_{ik}e_k$$

where $c_{ik} = \sum_j a_{ij}b_{jk}$. This shows that the matrix of $\eta\rho$ is $(a)(b)$. We have therefore proved that $\eta \to (a)$ is an isomorphism of $\mathfrak{E}$ into I_n.

We shall show finally that our mapping is onto I_n. Thus let (a) be any matrix in I_n and let $f_i = \sum_j a_{ij}e_j$. We define a mapping of $\mathfrak{G}$ into itself by stipulating that $\Sigma m_i e_i \to \Sigma m_i f_i$. Then if $\Sigma m_i' e_i$ is a second element of $\mathfrak{G}$, $\Sigma m_i e_i + \Sigma m_i' e_i = \Sigma(m_i + m_i')e_i$ and this element is mapped into

$$\Sigma(m_i + m_i')f_i = \Sigma m_i f_i + \Sigma m_i' f_i.$$

Hence $\Sigma m_i e_i \to \Sigma m_i f_i$ is an endomorphism η. Since $e_i\eta = f_i = \Sigma a_{ij}e_j$, the matrix of η is the given matrix (a). Thus we have established an isomorphism of $\mathfrak{E}$ onto I_n.

We can use the result which we have just derived to determine the group of automorphisms of $\mathfrak{G}$. It is clear that if $\mathfrak{G}$ is any commutative group, then the group of automorphisms $\mathfrak{A}$ of $\mathfrak{G}$ coincides with the group of units in the ring $\mathfrak{E}$. Also it is evident that if we have an isomorphism of one ring onto a second one, then the group of units of the first is mapped onto the group of units of the second. It follows that we can determine the group of automorphisms of the group $\mathfrak{G}$ of integral vectors by determining the group of units of the matrix ring I_n. Now we know that a matrix $(a) \,\epsilon\, I_n$ is a unit in I_n if and only if det $(a) = \pm 1$. This result in combination with the above discussion shows that the automorphisms of $\mathfrak{G}$ have the form $\Sigma m_i e_i \to \Sigma m_i f_i$ where $f_i = \Sigma a_{ij}e_j$ and det $(a) = \pm 1$.

EXERCISES

1. Determine the ring of endomorphisms and the group of automorphisms of a cyclic group of order n.

2. Let $\mathfrak{G}$ be an arbitrary group and let $\mathfrak{M}$ be the complete set of mappings of $\mathfrak{G}$ into itself. If $\eta, \rho \in \mathfrak{M}$, define $\eta\rho$ to be the resultant and $\eta + \rho$ by $g(\eta + \rho) = (g\eta)(g\rho)$. Investigate the set $\mathfrak{M}$ relative to these two compositions.

14. The multiplications of a ring. We suppose now that $\mathfrak{A}$ is any ring. If a is a fixed element of $\mathfrak{A}$, we define the *right multiplication* a_r to be the mapping $x \to xa$ of $\mathfrak{A}$ into itself. This mapping is an endomorphism of the additive group $\mathfrak{A},+$ of $\mathfrak{A}$ since

$$(27) \qquad (x + y)a_r = (x + y)a = xa + ya = xa_r + ya_r.$$

Next we note that

$$x(a + b)_r = x(a + b) = xa + xb = xa_r + xb_r = x(a_r + b_r)$$

and

$$x(ab)_r = x(ab) = (xa)b = (xa_r)b_r = x(a_r b_r).$$

Hence we have the relations

$$(28) \qquad \begin{aligned} (a + b)_r &= a_r + b_r \\ (ab)_r &= a_r b_r. \end{aligned}$$

These show that the correspondence $a \to a_r$ is a homomorphism of the ring $\mathfrak{A}$ into the ring $\mathfrak{C}$ of endomorphisms of $\mathfrak{A},+$. It follows, of course, that the set $\mathfrak{A}_r$ of the right multiplications is a subring of $\mathfrak{C}$. We shall call this *the ring of right multiplications* of the ring $\mathfrak{A}$.

The kernel of the homomorphism $a \to a_r$ is the ideal $\mathfrak{Z}_r$ of elements z such that $xz = 0$ for all x. We call this ideal the *right annihilator* of the ring $\mathfrak{A}$. If $\mathfrak{Z}_r = 0$, we know that $a \to a_r$ is an isomorphism. In particular we note that in the important case in which $\mathfrak{A}$ has an identity, $\mathfrak{Z}_r = 0$; for, if $1z = 0$, then $z = 0$. As a consequence we have proved the following fundamental

Theorem 10. *Any ring with an identity is isomorphic to a ring of endomorphisms.*

---

* We shall prove in the next chapter (p. 84) that this result is also valid for rings without identities.

A similar discussion applies to the *left multiplications* a_l defined by $xa_l = ax$. These mappings are endomorphisms and we have the rules

$$(29) \qquad (a + b)_l = a_l + b_l, \quad (ab)_l = b_l a_l.$$

It follows that $a \to a_l$ is an anti-homomorphism (cf. ex. 5, p. 74) of $\mathfrak{A}$ into $\mathfrak{E}$. Hence the image set, that is, the set $\mathfrak{A}_l$ of left multiplications, is a subring of $\mathfrak{E}$. The kernel of the anti-homomorphism $a \to a_l$ is the ideal $\mathfrak{Z}_l$ of left annihilators of the ring $\mathfrak{A}$. If $\mathfrak{A}$ has an identity, $\mathfrak{Z}_l = 0$ and $a \to a_l$ is an anti-isomorphism.

We consider finally an important relation between left and right multiplications for rings with an identity. This is stated in

Theorem 11. *If $\mathfrak{A}$ is a ring with an identity, then any mapping in $\mathfrak{A}, +$ that commutes with all the left (right) multiplications is a right (left) multiplication.*

The proof of this theorem is identical with that of the corresponding group result given on p. 30.

Chapter III

EXTENSIONS OF RINGS AND FIELDS

A given ring may fail to have certain properties that are necessary for solving a particular problem. However, it may be possible to construct a larger ring that has the required properties. Thus, for example, there exist equations of the form $ax = b$, $a \neq 0$ that have no solutions in the domain of integers. The field of rational numbers is constructed for the purpose of insuring the solvability of equations of this type. The method used to construct this extension can be generalized so as to apply to any commutative integral domain. This type of extension is one of those that we consider in this chapter. Among others we define also rings of polynomials, field extensions and rings of functions. We derive some of the properties of these extensions and, in particular, we determine the algebraic structure of any field.

1. Imbedding of a ring in a ring with an identity. In the preceding chapter we have proved that any ring with an identity is isomorphic to a ring of endomorphisms. We shall now show that any ring $\mathfrak{A}$ is isomorphic to a subring $\mathfrak{A}'$ of a ring $\mathfrak{B}$ that has an identity. Since $\mathfrak{B}$ is isomorphic to a ring of endomorphisms, it will follow that $\mathfrak{A}'$ and hence $\mathfrak{A}$ is isomorphic to a ring of endomorphisms.

In general we shall say that a ring $\mathfrak{A}$ is *imbedded* in a ring $\mathfrak{B}$ if $\mathfrak{B}$ contains a subring $\mathfrak{A}'$ isomorphic to $\mathfrak{A}$. The ring $\mathfrak{B}$ is called an *extension* of $\mathfrak{A}$.

In order to construct an extension of $\mathfrak{A}$ that has an identity we let $\mathfrak{B}$ be the product set $I \times \mathfrak{A}$ of pairs (m,a) where m is an integer and a is in the given ring $\mathfrak{A}$. Two pairs (m,a) and (n,b) are re-

84

garded as equal if and only if $m = n$ and $a = b$. We define an addition composition in $\mathfrak{B}$ by

$$(1) \qquad (m,a) + (n,b) = (m + n, a + b).$$

It is easy to see that $\mathfrak{B}, +$ is a commutative group. The 0 element is $(0,0)$ and $-(m,a) = (-m,-a)$. We define multiplication in $\mathfrak{B}$ by

$$(2) \qquad (m,a)(n,b) = (mn, na + mb + ab)$$

where on the right-hand side na and mb denote respectively the nth multiple of a and the mth multiple of b. Now

$$((m,a)(n,b))(q,c) = ((mn)q, q(na) + q(mb) + q(ab)$$
$$+ (mn)c + (na)c + (mb)c + (ab)c)$$

and

$$(m,a)((n,b)(q,c)) = (m(nq), m(nc) + m(qb) + m(bc)$$
$$+ a(nq) + a(nc) + a(qb) + a(bc)).$$

Hence the properties of multiples, the commutative law of addition and the associative laws in $\mathfrak{A}$ and in I yield the associative law of multiplication in $\mathfrak{B}$. Also

$$(m,a)[(n,b) + (q,c)]$$
$$= (m,a)(n + q, b + c)$$
$$= (m(n + q), m(b + c) + (n + q)a + a(b + c))$$
$$= (mn + mq, mb + mc + na + qa + ab + ac)$$

and

$$(m,a)(n,b) + (m,a)(q,c)$$
$$= (mn, mb + na + ab) + (mq, mc + qa + ac)$$
$$= (mn + mq, mb + na + ab + mc + qa + ac).$$

Hence one of the distributive laws holds. In a similar manner we can verify the other distributive law. Hence the system that we have constructed is a ring.

Using (2) we see that the element $1 = (1,0)$ acts as the identity in $\mathfrak{B}$. We consider next the subset $\mathfrak{A}'$ of $\mathfrak{B}$ of elements of the form $(0,a)$. Since

$$(0,a) + (0,b) = (0,a + b), \quad 0 = (0,0),$$

$$-(0,a) = (0,-a) \quad \text{and} \quad (0,a)(0,b) = (0,ab),$$

$\mathfrak{A}'$ is a subring of $\mathfrak{B}$. Also it is clear that, if we set $a' = (0,a)$, then the correspondence $a \to a'$ is an isomorphism of $\mathfrak{A}$ onto $\mathfrak{A}'$. Thus $\mathfrak{A}$ is imbedded in $\mathfrak{B}$, a ring with an identity. This proves the following

Theorem 1. *Any ring can be imbedded in a ring with an identity.*

We note also that the ring of integers is imbedded in the ring $\mathfrak{B}$ since the mapping $m \to (m,0)$ is an isomorphism of I onto a subring I' of $\mathfrak{B}$. We now simplify our notation by writing m for $(m,0)$ and a for $(0,a)$, I for I' and $\mathfrak{A}$ for $\mathfrak{A}'$. Using these notations, we have the relations

$$\mathfrak{B} = I + \mathfrak{A}, \quad I \cap \mathfrak{A} = 0.$$

Also it is clear that $\mathfrak{A}$ is an ideal in $\mathfrak{B}$.

Remarks. In certain situations the extension $\mathfrak{B}$ is not the best extension of $\mathfrak{A}$ to a ring with an identity element. In the first place, if $\mathfrak{A}$ has an identity e to begin with, then the element $z = 1 - e$ has the property $za = 0 = az$ for all a in $\mathfrak{A}$. Hence in this case it is not worthwhile to introduce the ring $\mathfrak{B}$. Next, we note that the characteristic of $\mathfrak{B}$ may be different from that of $\mathfrak{A}$. This will be the case if the characteristic of $\mathfrak{A}$ is $m \neq 0$; for $\mathfrak{B} \supseteq I$ and hence $\mathfrak{B}$ has characteristic 0. However, it is easy to modify the construction to obtain an extension with an identity that has the same characteristic as $\mathfrak{A}$. This is indicated in exercise 1 below. Another objection to the construction that we have given is that, if $\mathfrak{A}$ is an integral domain, $\mathfrak{B}$ may not be an integral domain. For instance, if $\mathfrak{A}$ is the ring of even integers, then the element $(2,-2)$ of $\mathfrak{B}$ has the property $(2,-2)(0,2m) = 0$. This difficulty can be overcome, too, and we can prove that any integral domain can be imbedded in an integral domain with an identity. Exercises 2–4 are designed to establish this result.

EXERCISES

1. Let $\mathfrak{A}$ be a ring for which there exists a positive integer m such that $ma = 0$ for all a. Let $\mathfrak{C}$ denote the set of pairs $(\bar{n},a)$ where $\bar{n} = n + (m)$ is in the ring $I/(m)$. Define equality as in the ring $\mathfrak{B}$, addition by $(\bar{n},a) + (\bar{q},b) = (\bar{n} + \bar{q}, a + b)$ and multiplication by

$$(\bar{n},a)(\bar{q},b) = (\bar{n}\bar{q},\, nb + qa + ab)$$

Show that multiplication is single-valued and that $\mathfrak{C}$ is a ring with an identity which is an extension of $\mathfrak{A}$ and that $mc = 0$ for all $c \,\varepsilon\, \mathfrak{C}$.

2. Let $\mathfrak{A}$ be an integral domain that contains elements a and $b \neq 0$ such that $ab + mb = 0$ for some integer m. Prove that $ca + mc = 0 = ac + mc$ for all c in $\mathfrak{A}$.

3. Let $\mathfrak{A}$ be an integral domain and let $\mathfrak{B}$ be the ring constructed in the text. Show that the totality $\mathfrak{Z}$ of elements z in $\mathfrak{B}$ such that $za = 0$ for all a in $\mathfrak{A}$ is an ideal and that $\mathfrak{B}/\mathfrak{Z}$ is an integral domain with an identity.

4. Prove that the set $\overline{\mathfrak{A}}$ of cosets of the form $a + \mathfrak{Z}$, a in $\mathfrak{A}$, is a subring of $\mathfrak{B}/\mathfrak{Z}$ isomorphic to $\mathfrak{A}$. Hence $\mathfrak{A}$ is imbedded in $\mathfrak{B}/\mathfrak{Z}$.

2. Field of fractions of a commutative integral domain.

We shall now show that any commutative integral domain can be imbedded in a field. The construction which we shall give—well known for the ring of integers—can best be understood by studying the relation between a subring of a field and the subfield generated by the subring.

Hence let $\mathfrak{F}$ be a field and let $\mathfrak{A}$ be a subring $\neq 0$ of $\mathfrak{F}$. We say that $\mathfrak{A}$ is a *subfield* of $\mathfrak{F}$ if the system $\mathfrak{A},+,\cdot$ is a field. It is immediate that a subset $\mathfrak{A}$ of a field $\mathfrak{F}$ determines a subfield if and only if (1) $\mathfrak{A},+$ is a subgroup of the additive group. (2) $\mathfrak{A}$ contains elements $\neq 0$, and if $\mathfrak{A}^*$ denotes the totality of these elements, then $\mathfrak{A}^*,\cdot$ is a subgroup of the multiplicative group of non-zero elements of $\mathfrak{F}$. If we recall the conditions that a subset of a group determines a subgroup, we see that $\mathfrak{A}$ determines a subfield if and only if

1′. If $a, b \,\varepsilon\, \mathfrak{A}$, then $a + b \,\varepsilon\, \mathfrak{A}$. $0 \,\varepsilon\, \mathfrak{A}$. If $a \,\varepsilon\, \mathfrak{A}$, then $-a \,\varepsilon\, \mathfrak{A}$.

2′. $1 \,\varepsilon\, \mathfrak{A}$. If a and b are non-zero elements of $\mathfrak{A}$, then ab and $a^{-1} \,\varepsilon\, \mathfrak{A}$.

It is clear from 1′ and 2′ that the intersection of any collection of subfields of a field is again a subfield. If S is any subset of $\mathfrak{F}$, then the intersection of all subfields of $\mathfrak{F}$ that contain S is called the *smallest subfield of $\mathfrak{F}$ containing S* or the *subfield of $\mathfrak{F}$ generated by S*. We now make the important observation that, if $S = \mathfrak{A}$ is a subring $\neq 0$ of $\mathfrak{F}$, then the subfield $\mathfrak{G}$ generated by $\mathfrak{A}$ coincides with the set $\{ab^{-1}\}$ of elements of the form ab^{-1}, a and b in $\mathfrak{A}$. First, it is clear that $\mathfrak{G} \supseteq \{ab^{-1}\}$. Also we have the following equations:

$$ab^{-1} + cd^{-1} = adb^{-1}d^{-1} + cbb^{-1}d^{-1} = (ad + cb)(bd)^{-1}$$

$$0 = 0b^{-1}$$

$$-ab^{-1} = (-a)b^{-1}$$

$$(ab^{-1})(cd^{-1}) = acb^{-1}d^{-1} = (ac)(bd)^{-1}$$

$$1 = aa^{-1} \quad (a \neq 0)$$

$$(ab^{-1})^{-1} = a^{-1}b \quad (a \neq 0),$$

and these show that the set $\{ab^{-1}\}$ determines a subfield. Since any a in $\mathfrak{A}$ has the form

$$a = (ab)b^{-1},$$

$\mathfrak{A} \subseteq \{ab^{-1}\}$. Hence the set $\{ab^{-1}\}$ is a subfield of $\mathfrak{F}$ containing $\mathfrak{A}$. Since $\mathfrak{G} \supseteq \{ab^{-1}\}$ this implies that $\mathfrak{G} = \{ab^{-1}\}$.

If $\mathfrak{F} = \mathfrak{G}$, then we shall say that $\mathfrak{F}$ is a *minimal field* containing $\mathfrak{A}$. In this case we see that every element of $\mathfrak{F}$ has the form ab^{-1}, a and b in $\mathfrak{A}$.

Suppose now that $\mathfrak{A}$ is any commutative integral domain $\neq 0$. We wish to extend $\mathfrak{A}$ to a field. The foregoing remarks indicate that the elements of a minimal field extension of $\mathfrak{A}$ are to be obtained from the pairs (a,b), $b \neq 0$ and a in $\mathfrak{A}$. We have in mind that (a,b) is to play the role of ab^{-1}. Hence we adopt the following procedure.

Let $\mathfrak{B}$ be the totality of pairs (a,b), $b \neq 0$ and a in $\mathfrak{A}$. We introduce a relation $\sim$ in $\mathfrak{B}$ by defining $(a,b) \sim (c,d)$ if $ad = bc$. Then $(a,b) \sim (a,b)$ since $ab = ba$ and, if $(a,b) \sim (c,d)$, $ad = bc$ so that $cb = da$ and $(c,d) \sim (a,b)$. Finally if $(a,b) \sim (c,d)$ and $(c,d) \sim (e,f)$, then $ad = bc$ and $cf = de$. Hence $adf = bcf = bde$. Since $d \neq 0$ and $\mathfrak{A}$ is commutative, d may be cancelled to give $af = be$. Hence $(a,b) \sim (e,f)$. We have therefore proved that the relation $\sim$ is an equivalence relation in $\mathfrak{B}$. We shall call the equivalence class determined by (a,b) the *fraction a/b*. Thus we have the rule

$$a/b = c/d \quad \text{if and only if} \quad ad = bc.$$

We shall now introduce addition and multiplication compositions in the set $\mathfrak{F}$ of fractions. We note first that, if a/b and c/d

are any two fractions, then $bd \neq 0$ and we can form the fraction $(ad + bc)/bd$.   Moreover, if $a/b = a'/b'$ and $c/d = c'/d'$, then

$$(3) \qquad (ad + bc)/bd = (a'd' + b'c')/b'd'.$$

Thus, by assumption, $ab' = ba'$ and $cd' = dc'$.   Hence

$$ab'dd' = ba'dd' \quad \text{and} \quad cd'bb' = dc'bb'$$

so that

$$ab'dd' + cd'bb' = ba'dd' + dc'bb'$$

or

$$(ad + bc)b'd' = (a'd' + b'c')bd,$$

and this is equivalent to (3).   It is now clear that the addition composition defined by

$$(4) \qquad a/b + c/d = (ad + bc)/bd$$

is a single-valued composition in $\mathfrak{F}$.   In a similar manner we see that, if a/b and c/d are fractions, then ac/bd is a fraction. If $a/b = a'/b'$ and $c/d = c'/d'$, then $ac/bd = a'c'/b'd'$.   Hence

$$(5) \qquad (a/b)(c/d) = ac/bd$$

defines a single-valued multiplication.

It can also be verified directly that $\mathfrak{F}$ with the compositions (4) and (5) is a commutative ring.   We leave this verification to the reader.   It will be observed that $0/b = 0/d$ is the 0 of $\mathfrak{F}$ and that the negative of a/b is $(-a)/b = a/(-b)$.   The ring $\mathfrak{F}$ has an identity; for $b/b = d/d$ for any $b \neq 0$ and $d \neq 0$ and $(a/b)(b/b) = ab/b^2 = a/b$.   Hence $b/b = 1$.   If $a/b \neq 0$ then $a \neq 0$.   Hence b/a is a fraction.   Since $(a/b)(b/a) = ab/ba = 1$, $b/a = (a/b)^{-1}$.   This shows that every element $\neq 0$ in $\mathfrak{F}$ is a unit.   Hence $\mathfrak{F}$ is a field.

We now associate with the element a of $\mathfrak{A}$ the fraction ab/b where b is any element $\neq 0$ in $\mathfrak{A}$.   This correspondence is single-valued since $ab/b = ad/d$ for any $d \neq 0$.   We denote ab/b by $\bar{a}$. Then

$$\begin{aligned}
\overline{a + a'} &= (a + a')b/b = (a + a')b^2/b^2 = (ab^2 + a'b^2)/b^2 \\
&= ab/b + a'b/b \\
&= \bar{a} + \overline{a'}
\end{aligned}$$

and

$$\overline{aa'} = aa'b/b = aa'b^2/b^2 = (ab/b)(a'b/b)$$
$$= \bar{a}\overline{a'},$$

so that $a \to \bar{a}$ is a homomorphism. Also we can verify directly that this mapping is 1–1. Hence the set $\overline{\mathfrak{A}}$ of elements $\bar{a}$ determines a subring of $\mathfrak{F}$ isomorphic to $\mathfrak{A}$. We have therefore proved the following fundamental imbedding theorem.

Theorem 2. *Any commutative integral domain ($\neq 0$) can be imbedded in a field.*

We shall now note that $\mathfrak{F}$ is a minimal field containing the image $\overline{\mathfrak{A}}$ of $\mathfrak{A}$. This is clear since any a/b of $\mathfrak{F}$ can be written in the form $a/b = (ab/b)(b/b^2) = (ab/b)(b^2/b)^{-1} = \bar{a}\bar{b}^{-1}$.

If $\mathfrak{A} = I$ the ring of integers, then the fractions are called *rational numbers*. We denote the field of rational numbers by R_0 in the sequel.

EXERCISES

1. Show that, if $\mathfrak{A}$ is a field, then $\mathfrak{F} = \overline{\mathfrak{A}}$.
2. Prove that any commutative semi-group that satisfies the cancellation law can be imbedded in a group.

GENERALIZATIONS. (1) The method that we have just used can be extended to prove that any commutative ring $\mathfrak{A}$ that contains a non-vacuous set S of elements that are not zero-divisors can be imbedded in a ring with an identity in which the elements of S are units.

We note first that, if $s_1 s_2$ is a zero-divisor, then either s_1 or s_2 is a zero-divisor. Hence the sub-semigroup V of the multiplicative semi-group of $\mathfrak{A}$ generated by the given set S contains no zero-divisors. We consider now the set $\mathfrak{A} \times V$ of pairs (a,v) a in $\mathfrak{A}$, v in V, and we introduce the relation $(a,v) \sim (a',v')$ if $av' = a'v$. This is an equivalence relation since V contains no zero-divisors. Let $\mathfrak{F}_S = \mathfrak{F}_V$ be the set of equivalence classes a/v determined by this relation. Addition and multiplication are defined as before. We obtain in this way a ring that contains a subring $\overline{\mathfrak{A}} \cong \mathfrak{A}$. The elements of $\overline{\mathfrak{A}}$ are the classes $\bar{a} = av/v$. The ring $\mathfrak{F}_S$ is commutative and has the identity v/v. If $s \varepsilon S$, the corresponding element $\bar{s} = sv/v$ is a unit in $\mathfrak{F}_S$; its inverse is v/sv.

(2) There is an important class of non-commutative integral domains that can be imbedded in division rings. These are the domains that have the *common multiple property*, that is, any pair of non-zero elements a,b in the domain has a common right (left) multiple $m = ab' = ba' \neq 0$ ($\bar{m} = \bar{b}a = \bar{a}b$). The imbedding problem for integral domains of this type was first solved by O. Ore. The construction is similar to the one we have used in the commutative case. We refer the reader to Ore's paper for the details.*

We note finally that it has been proved by A. Malcev that there exist non-commutative integral domains that cannot be imbedded in division rings.†

3. Uniqueness of the field of fractions. Let $\mathfrak{A}$ be a commutative integral domain and let $\mathfrak{F}$ be its field of fractions. We shall now identify $\mathfrak{A}$ with the subring $\bar{\mathfrak{A}}$ of elements $\bar{a} = ab/b$. Thus we shall write $\mathfrak{A}$ for $\bar{\mathfrak{A}}$, a for $\bar{a}$. Then we know that the subfield of $\mathfrak{F}$ generated by $\mathfrak{A}$ is $\mathfrak{F}$ itself. We shall now prove that any two fields that bear this relation to $\mathfrak{A}$ are isomorphic. More precisely, we have the following

Theorem 3. *Let $\mathfrak{A}_i$, $i = 1, 2$ be a subring $\neq 0$ of the field $\mathfrak{F}_i$ and suppose that $\mathfrak{F}_i$ is the smallest subfield of $\mathfrak{F}_i$ containing $\mathfrak{A}_i$. Then if σ is an isomorphism of $\mathfrak{A}_1$ onto $\mathfrak{A}_2$, σ can be extended in one and only one way to an isomorphism of $\mathfrak{F}_1$ onto $\mathfrak{F}_2$.*

By an *extension of a mapping* of a set to a mapping of a larger set we mean a mapping of the larger set that has the same effect as the original mapping on the elements of the given subset. Then we have to find an isomorphism Σ of $\mathfrak{F}_1$ onto $\mathfrak{F}_2$ such that $a_1^{\Sigma} = a_1^{\sigma}$ for all $a_1 \, \varepsilon \, \mathfrak{A}_1$. We shall now verify that the mapping

$$(6) \qquad a_1 b_1^{-1} \rightarrow a_1^{\sigma}(b_1^{\sigma})^{-1}$$

$b_1 \neq 0$ in $\mathfrak{A}_1$ has the required properties. In the first place, since $\mathfrak{F}_1$ is minimal for $\mathfrak{A}_1$, any element of $\mathfrak{F}_1$ has the form $a_1 b_1^{-1}$. Hence (6) is defined for the whole of $\mathfrak{F}_1$. We note next that (6)

---

* O. Ore, *Linear equations in non-commutative fields*, Annals of Mathematics, Vol. 32 (1931), pp. 463–477.

† A. Malcev, *On the immersion of an algebraic ring into a field*, Mathematische Annalen, Vol. 113 (1937), 686–691.

is single-valued; for suppose that $a_1 b_1{}^{-1} = c_1 d_1{}^{-1}$. Then $a_1 d_1 = c_1 b_1$ and $a_1{}^\sigma d_1{}^\sigma = c_1{}^\sigma b_1{}^\sigma$. Hence $a_1{}^\sigma (b_1{}^\sigma)^{-1} = c_1{}^\sigma (d_1{}^\sigma)^{-1}$ as required. In a similar manner we see that, if $a_1{}^\sigma (b_1{}^\sigma)^{-1} = c_1{}^\sigma (d_1{}^\sigma)^{-1}$, then $a_1 b_1{}^{-1} = c_1 d_1{}^{-1}$; hence the mapping is 1–1. If $a_2 b_2{}^{-1}$ is any element of $\mathfrak{F}_2$ we can find an a_1 such that $a_1{}^\sigma = a_2$ and a b_1 such that $b_1{}^\sigma = b_2$. Then $a_2 b_2{}^{-1} = a_1{}^\sigma (b_1{}^\sigma)^{-1}$ is an image. Hence our mapping is a mapping of $\mathfrak{F}_1$ *onto* $\mathfrak{F}_2$. Finally we note that

$$a_1 b_1{}^{-1} + c_1 d_1{}^{-1}$$

$$= (a_1 d_1 + c_1 b_1)(b_1 d_1)^{-1} \rightarrow (a_1 d_1 + c_1 b_1)^\sigma ((b_1 d_1)^\sigma)^{-1}$$

$$= (a_1{}^\sigma d_1{}^\sigma + c_1{}^\sigma b_1{}^\sigma)(b_1{}^\sigma d_1{}^\sigma)^{-1}$$

$$= a_1{}^\sigma (b_1{}^\sigma)^{-1} + c_1{}^\sigma (d_1{}^\sigma)^{-1}$$

and

$$(a_1 b_1{}^{-1})(c_1 d_1{}^{-1}) = a_1 c_1 (b_1 d_1)^{-1} \rightarrow (a_1 c_1)^\sigma ((b_1 d_1)^\sigma)^{-1}$$

$$= (a_1{}^\sigma c_1{}^\sigma)(b_1{}^\sigma d_1{}^\sigma)^{-1}$$

$$= (a_1{}^\sigma (b_1{}^\sigma)^{-1})(c_1{}^\sigma (d_1{}^\sigma)^{-1}).$$

Hence we have an isomorphism of $\mathfrak{F}_1$ onto $\mathfrak{F}_2$. This isomorphism is an extension of σ since it maps $a_1 = (a_1 b_1) b_1{}^{-1}$ into

$$(a_1 b_1)^\sigma (b_1{}^\sigma)^{-1} = a_1{}^\sigma b_1{}^\sigma (b_1{}^\sigma)^{-1} = a_1{}^\sigma.$$

Suppose now that Σ is any isomorphism of $\mathfrak{F}_1$ onto $\mathfrak{F}_2$ that coincides with σ in $\mathfrak{A}_1$. Then $(a_1 b_1{}^{-1})^\Sigma = a_1{}^\Sigma (b_1{}^{-1})^\Sigma = a_1{}^\Sigma (b_1{}^\Sigma)^{-1} = a_1{}^\sigma (b_1{}^\sigma)^{-1}$. Hence Σ is the mapping (6). This shows that the extension of σ to an isomorphism of $\mathfrak{F}_1$ onto $\mathfrak{F}_2$ is uniquely determined. The theorem is therefore completely proved.

4. Polynomial rings. One is often interested in studying a ring $\mathfrak{B}$ relative to a specified subring $\mathfrak{A}$. As we shall see, this idea is particularly fruitful in the theory of fields. A natural problem in this connection is the determination of the structure of a subring $\mathfrak{A}[u]$ generated by $\mathfrak{A}$ and one additional element $u \, \varepsilon \, \mathfrak{B}$. To simplify this problem we shall assume that (1) $\mathfrak{B}$ has an identity 1, (2) 1 is in $\mathfrak{A}$, (3) $ua = au$ for all a in $\mathfrak{A}$. Evidently any element of the form

$$(7) \qquad\qquad a_0 + a_1 u + a_2 u^2 + \cdots + a_n u^n$$

where the $a_i \, \varepsilon \, \mathfrak{A}$ is in $\mathfrak{A}[u]$. We shall call an element of this form a *polynomial* in u with *coefficients* a_i in $\mathfrak{A}$.

If $b_0 + b_1 u + b_2 u^2 + \cdots + b_m u^m$ is a second polynomial in u and $n \geq m$, then

$$(8) \quad (a_0 + a_1 u + a_2 u^2 + \cdots + a_n u^n)$$
$$+ (b_0 + b_1 u + b_2 u^2 + \cdots + b_m u^m)$$
$$= (a_0 + b_0) + (a_1 + b_1)u + \cdots$$
$$+ (a_m + b_m)u^m + a_{m+1} u^{m+1} + \cdots + a_n u^n.$$

Also 0 is a polynomial and the negative of $\sum_0^n a_i u^i$ is the polynomial $\Sigma(-a_i)u^i$. Finally, since $(a_i u^i)(b_j u^j) = a_i b_j u^{i+j}$,

$$(9) \quad (a_0 + a_1 u + a_2 u^2 + \cdots + a_n u^n)(b_0 + b_1 u + \cdots + b_m u^m)$$
$$= p_0 + p_1 u + \cdots + p_{n+m} u^{n+m}$$

where

$$(10) \qquad p_i = \sum_{j=0}^{i} a_j b_{i-j} \equiv \sum_{j+k=i} a_j b_k.$$

Hence the totality of polynomials is a subring of $\mathfrak{B}$. Clearly this subring includes $\mathfrak{A}$ and, since $\mathfrak{A}$ contains 1, $u = 1u$ is a polynomial. It follows that the ring $\mathfrak{A}[u]$ generated by $\mathfrak{A}$ and by u is just the set of polynomials in u with coefficients in $\mathfrak{A}$.

A particularly simple situation is obtained when the element u is *transcendental* relative to $\mathfrak{A}$. By this we mean that a polynomial relation

$$d_0 + d_1 u + d_2 u^2 + \cdots + d_m u^m = 0,$$

d_i in $\mathfrak{A}$ can hold only if all the $d_i = 0$. In this case the two polynomials $\sum_0^n a_i u^i$ and $\sum_0^m b_j u^j$ are equal only if the corresponding coefficients a_i and b_i are equal for all i; for if $n \geq m$ and $\Sigma \, a_i u^i = \Sigma b_j u^j$, then

$$(a_0 - b_0) + (a_1 - b_1)u + \cdots + (a_m - b_m)u^m + a_{m+1} u^{m+1} + \cdots$$
$$+ a_n u^n = 0.$$

Hence $a_j = b_j, \, j = 1, 2, \cdots, m$ and $a_{m+1} = \cdots = a_n = 0$.

If u is not transcendental, we say that u is an *algebraic element* relative to the subring $\mathfrak{A}$. In order to determine the structure of polynomial rings it is important to have available rings of the form $\mathfrak{A}[x]$ where x is transcendental. In a polynomial extension by a transcendental element the polynomial (7) determines a unique sequence $(a_0, a_1, \cdots)$ with the property that $a_i = 0$ for sufficiently large i. Hence it is natural to adopt the following procedure for constructing $\mathfrak{A}[x]$.

Let $\mathfrak{A}$ be a given ring with an identity and let $\mathfrak{B}$ be the totality of infinite sequences

$$(a_0, a_1, a_2, \cdots)$$

that have only a finite number of non-zero terms a_i. Elements of $\mathfrak{B}$ are regarded as equal if and only if $a_i = b_i$ for all i. Addition in $\mathfrak{B}$ is defined by

$$(11) \quad (a_0, a_1, a_2, \cdots) + (b_0, b_1, b_2, \cdots)$$
$$= (a_0 + b_0, a_1 + b_1, a_2 + b_2, \cdots).$$

The result given in the right-hand side is a member of $\mathfrak{B}$ since the terms in the sequence are all 0 from a certain point on. It is immediate that $\mathfrak{B}$ is a commutative group relative to this addition. The $0 = (0, 0, \cdots)$ and $-(a_0, a_1, \cdots) = (-a_0, -a_1, \cdots)$. We define multiplication in $\mathfrak{B}$ by

$$(12) \quad (a_0, a_1, a_2, \cdots)(b_0, b_1, b_2, \cdots) = (p_0, p_1, p_2, \cdots)$$

where p_i is given by (10). If $a_i = 0$ for $i > n$ and $b_j = 0$ for $j > m$, then $p_k = 0$ for $k > m + n$. Hence (12) gives an element of $\mathfrak{B}$.

If $a = (a_0, a_1, \cdots)$, $b = (b_0, b_1, \cdots)$ and $c = (c_0, c_1, \cdots)$, then the term with subscript i in $(ab)c$ is

$$\sum_{m+l=i} \left(\sum_{j+k=m} a_j b_k \right) c_l = \sum_{j+k+l=i} a_j b_k c_l.$$

Similarly the corresponding term of $a(bc)$ is

$$\sum_{m+j=i} a_j \left(\sum_{k+l=m} b_k c_l \right) = \sum_{j+k+l=i} a_j b_k c_l.$$

Hence $(ab)c = a(bc)$. Similarly we can verify the distributive laws. Hence the system $\mathfrak{B}, +, \cdot$ is a ring.

The subset $\mathfrak{A}'$ of elements

$$a' = (a, 0, 0, \cdots)$$

is a subring of $\mathfrak{B}$ isomorphic under the correspondence $a \to a'$ with $\mathfrak{A}$. Thus $\mathfrak{A}$ is imbedded in $\mathfrak{B}$. The element $1' = (1, 0, \cdots)$ of $\mathfrak{A}'$ acts as an identity in $\mathfrak{B}$. Now let x denote the element $(0, 1, 0, 0, \cdots)$. Then

$$x^k = (0, 0, \cdots, \overset{k+1}{0, 1}, 0, \cdots)$$

and

$$a'x^k = (0, 0, \cdots, \overset{k+1}{0, a}, 0, \cdots) = x^k a'.$$

Hence x commutes with every $a' \, \varepsilon \, \mathfrak{A}'$ and the general element $(a_0, a_1, \cdots, a_n, 0, 0, \cdots)$ can be written as

$$(13) \qquad a_0' + a_1'x + a_2'x^2 + \cdots + a_n'x^n.$$

Thus $\mathfrak{B} = \mathfrak{A}'[x]$. If (13) is 0, then $(a_0, a_1, \cdots) = 0$. Hence all the a_i and therefore all the $a_i' = 0$. This shows that x is transcendental relative to $\mathfrak{A}'$.

It will now be well to replace the ring $\mathfrak{A}$ by the isomorphic ring $\mathfrak{A}'$ and to denote the latter by $\mathfrak{A}$. We shall also write a for the element a'. Then $\mathfrak{B} = \mathfrak{A}[x]$ and x is transcendental relative to $\mathfrak{A}$ as we required.

EXERCISES

1. Let $\mathfrak{B}^*$ be the complete set of sequences $(a_0, a_1, a_2, \cdots)$ with $a_i \, \varepsilon \, \mathfrak{A}$. Define equality, addition and multiplication as for the ring $\mathfrak{B}$. Prove that $\mathfrak{B}^*$ is a ring. This ring is called the ring of *formal power series* over $\mathfrak{A}$ and will be denoted as $\mathfrak{A} < x >$ in the sequel.

2. Let S be any semi-group and let $\mathfrak{A}$ be any ring. Denote by $\mathfrak{B}$ the set of functions $a(s)$ defined on S and having values in $\mathfrak{A}$ such that $a(s) = 0$ for all but a finite number of s. Define addition and multiplication in $\mathfrak{B}$ by

$$(a + b)(s) = a(s) + b(s)$$

$$(ab)(s) = \sum_{tu=s} a(t)b(u).$$

Show that $\mathfrak{B}$ is a ring. We shall call $\mathfrak{B}$ a *semi-group ring*.

3. Show that the semi-group ring determined by the semi-group of non-negative integers with addition as composition is the ring $\mathfrak{A}[x]$ constructed above.

5. Structure of polynomial rings. Let $\mathfrak{A}_1[x]$ be a polynomial ring in an element x that is transcendental over the base ring $\mathfrak{A}_1$ and let $\mathfrak{A}_2[u]$ be an arbitrary polynomial ring such that $\mathfrak{A}_2$ is a homomorphic image of $\mathfrak{A}_1$. As before, we assume that our rings contain identities and that the elements x, u commute with the displayed coefficient rings. Let σ be a definite homomorphism of $\mathfrak{A}_1$ onto $\mathfrak{A}_2$. Then we shall show that this homomorphism can be extended in one and only one way to a homomorphism of $\mathfrak{A}_1[x]$ onto $\mathfrak{A}_2[u]$ mapping x into u.

Since x is transcendental, an element of $\mathfrak{A}_1[x]$ can be written in one and only one way in the form

$$a_0 + a_1 x + \cdots + a_n x^n, \quad a_i \text{ in } \mathfrak{A}_1.$$

We now denote this element as $f(x)$ and we define

$$f^\sigma(u) = a_0^\sigma + a_1^\sigma u + \cdots + a_n^\sigma u^n, \quad a_i^\sigma \text{ in } \mathfrak{A}_2.$$

It is clear that the rule $f(x) \to f^\sigma(u)$ defines a single-valued mapping of $\mathfrak{A}_1[x]$ onto $\mathfrak{A}_2[u]$. If $g(x) = \Sigma b_i x^i$, then $f(x) + g(x) = \Sigma(a_i + b_i)x^i$ and this element is mapped into

$$\Sigma(a_i + b_i)^\sigma u^i = \Sigma(a_i^\sigma + b_i^\sigma)u^i$$

$$= \Sigma a_i^\sigma u^i + \Sigma b_i^\sigma u^i.$$

Also

$$f(x)g(x) = a_0 b_0 + (a_0 b_1 + a_1 b_0)x + (a_0 b_2 + a_1 b_1 + a_2 b_0)x^2 + \cdots$$

$$\to (a_0 b_0)^\sigma + (a_0 b_1 + a_1 b_0)^\sigma u$$

$$+ (a_0 b_2 + a_1 b_1 + a_2 b_0)^\sigma u^2 + \cdots$$

$$= a_0^\sigma b_0^\sigma + (a_0^\sigma b_1^\sigma + a_1^\sigma b_0^\sigma)u$$

$$+ (a_0^\sigma b_2^\sigma + a_1^\sigma b_1^\sigma + a_2^\sigma b_0^\sigma)u^2 + \cdots$$

$$= (\Sigma a_i^\sigma u^i)(\Sigma b_i^\sigma u^i).$$

Hence our mapping is a homomorphism. Clearly, if $a \,\varepsilon\, \mathfrak{A}_1$, then $a \to a^\sigma$ in the new mapping. Moreover $x \to u$. Hence the mapping meets all of the requirements that we imposed.

Now let Σ be any homomorphism of $\mathfrak{A}_1[x]$ onto $\mathfrak{A}_2[u]$ that maps x into u and that coincides with σ on $\mathfrak{A}_1$. Then $(\Sigma a_i x^i)^\Sigma = \Sigma a_i^\Sigma u^i = \Sigma a_i^\sigma u^i$. Hence Σ coincides with the mapping that we have defined. This proves the uniqueness of the extension. We therefore have the following important homomorphism theorem.

Theorem 4. *Let $\mathfrak{A}_1[x]$ be a ring of polynomials in a transcendental element x and let $\mathfrak{A}_2[u]$ be a ring of polynomials in an arbitrary u. Suppose that σ is a homomorphism of $\mathfrak{A}_1$ onto $\mathfrak{A}_2$. Then σ can be extended in one and only one way to a homomorphism Σ of $\mathfrak{A}_1[x]$ onto $\mathfrak{A}_2[u]$ mapping x into u.*

If $\mathfrak{A} = \mathfrak{A}_1 = \mathfrak{A}_2$ and σ is the identity mapping, then this theorem shows that $\mathfrak{A}[u]$ for arbitrary u is a homomorphic image of $\mathfrak{A}[x]$, x transcendental. Hence by the fundamental theorem of homomorphism $\mathfrak{A}[u] \cong \mathfrak{A}[x]/\mathfrak{K}$, where $\mathfrak{K}$, the kernel of the homomorphism, is an ideal in $\mathfrak{A}[x]$. Since the homomorphism Σ is the identity mapping in $\mathfrak{A}$ it is clear that $\mathfrak{A} \cap \mathfrak{K} = 0$. Assume now that u, too, is transcendental. Then if $f(x)^\Sigma = 0, f(u) = 0$; hence $f(x) = 0$. This shows that $\mathfrak{K} = 0$. Hence Σ is an isomorphism. We therefore have the following

Theorem 5. *If x and y are transcendental over $\mathfrak{A}$, then $\mathfrak{A}[x]$ and $\mathfrak{A}[y]$ are isomorphic. Any ring of the form $\mathfrak{A}[u]$ is isomorphic to a difference ring $\mathfrak{A}[x]/\mathfrak{K}$ where x is transcendental and $\mathfrak{K}$ is an ideal in $\mathfrak{A}[x]$ such that $\mathfrak{K} \cap \mathfrak{A} = 0$.*

6. Properties of the ring $\mathfrak{A}[x]$. From now on x will denote a transcendental element over $\mathfrak{A}$. If $f(x)$ is a non-zero polynomial in $\mathfrak{A}[x]$, we can write $f(x) = a_0 + a_1 x + \cdots + a_n x^n$ with $a_n \neq 0$. We call a_n the *leading coefficient* of $f(x)$ and we call n the *degree* of $f(x)$. If $f(x) = 0$, we say that its degree is $-\infty$, and we adopt the usual conventions that $-\infty - \infty = -\infty, -\infty + n = -\infty$.

If a_n is not a left zero-divisor in $\mathfrak{A}$ and $g(x) = b_0 + b_1 x + \cdots + b_m x^m$ with $b_m \neq 0$, then

$$f(x)g(x) = a_0 b_0 + (a_0 b_1 + a_1 b_0)x + \cdots + a_n b_m x^{n+m}.$$

Since $a_n b_m \neq 0, f(x)g(x) \neq 0$ and this polynomial has the degree $m + n$. A similar result holds for $g(x)f(x)$ if a_n is not a right zero-divisor. In particular we see that *if $\mathfrak{A}$ is an integral domain then $\mathfrak{A}[x]$ is an integral domain.* Moreover, in this case we have the formula

$$(14) \qquad \deg f(x)g(x) = \deg f(x) + \deg g(x)$$

for all f and g. This has been proved above for the case $f \neq 0$ and $g \neq 0$, and it follows if either $f = 0$ or $g = 0$ by the conventions on $-\infty$. We note also the following useful result concerning the degree:

$$(15) \qquad \deg\,[f(x) + g(x)] \leq \max\,(\deg f(x),\, \deg g(x)).$$

The degree relation (14) enables us to determine the units in $\mathfrak{A}[x]$; for if $f(x)g(x) = 1$, the $\deg\,f(x) + \deg\,g(x) = 0$. Hence $\deg\,f(x) = 0 = \deg\,g(x)$. Thus $f(x) = a\,\varepsilon\,\mathfrak{A}$ and $g(x) = b\,\varepsilon\,\mathfrak{A}$. This proves that, *if $\mathfrak{A}$ is an integral domain, then the only units in $\mathfrak{A}[x]$ are the elements of $\mathfrak{A}$ that are units in $\mathfrak{A}$.* For example, if I is the ring of integers, the only units in $I[x]$ are the integers ± 1 and, if $\mathfrak{F}$ is a field, then the units of $\mathfrak{F}[x]$ are the non-zero elements of $\mathfrak{F}$.

We consider again the case of an arbitrary $\mathfrak{A}$ and we wish to establish a division process in $\mathfrak{A}[x]$. Let $g(x) = b_0 + b_1 x + \cdots + b_m x^m$ be any non-zero polynomial *whose leading coefficient b_m is a unit.* Suppose that $f(x)$ is arbitrary. Then we shall show that there exist polynomials $q_1(x)$ and $r_1(x)$ such that $\deg\,r_1(x) < \deg\,g(x)$ and

$$(16) \qquad\qquad f(x) = q_1(x)g(x) + r_1(x).$$

If $\deg\,f(x) < \deg\,g(x)$, we write $f(x) = q(x) \cdot 0 + f(x)$ to obtain the required representation. Assume now that $f(x) = a_0 + a_1 x + \cdots + a_n x^n$ has degree $n \geq m$. Also, using induction, we may assume that the result holds for polynomials f of degree $< n$. Let

$$f(x) - a_n b_m^{-1} x^{n-m} g(x) = f_1(x).$$

Then the terms $a_n x^n$ of maximum degree in $f(x)$ and in $a_n b_m^{-1} x^{n-m} g(x)$ cancel off so that $\deg f_1(x) < \deg f(x)$. Hence we may suppose that there exists a $q^*(x)$ and a $r_1(x)$ of degree less than m such that

$$f_1(x) = q^*(x)g(x) + r_1(x).$$

Then

$$f(x) = a_n b_m^{-1} x^{n-m} g(x) + q^*(x)g(x) + r_1(x)$$

$$= q_1(x)g(x) + r_1(x)$$

where $q_1(x) = a_n b_m^{-1} x^{n-m} + q^*(x)$ and $\deg\,r_1(x) < \deg\,g(x)$.

The "right-hand quotient" $q_1(x)$ and the "right-hand remainder" $r_1(x)$ are unique; for suppose that

$$f(x) = q_2(x)g(x) + r_2(x), \quad \deg r_2(x) < \deg g(x).$$

Then

$$[q_1(x) - q_2(x)]g(x) = r_2(x) - r_1(x).$$

The degree of the right-hand side is $<m$, while the degree of the left-hand side is either $-\infty$ or $\geq m$. Hence the common value must be $-\infty$ so that $r_2(x) - r_1(x) = 0$ and $q_1(x) - q_2(x) = 0$.

In a similar manner we can prove the existence and uniqueness of the "left-hand quotient" $q_2(x)$ and "left-hand remainder" $r_2(x)$ of degree $< \deg g(x)$ such that

$$f(x) = g(x)q_2(x) + r_2(x).$$

We consider now the special case in which $g(x) = x - c$, c in $\mathfrak{A}$. In order to obtain a formula for the remainder on division by $(x - c)$ we make use of the following identities:

$$(17) \quad x^k - c^k = (x^{k-1} + cx^{k-2} + c^2x^{k-3} + \cdots + c^{k-1})(x - c)$$

$$= (x - c)(x^{k-1} + cx^{k-2} + \cdots + c^{k-1}),$$

$k = 0, 1, 2, \cdots$. Here it is understood that, if $k = 0$, then the factor $\Sigma c^i x^{k-i-1} = 0$. We multiply (17) on the left by a_k and sum on k. This gives

$$f(x) - f_R(c) = q_1(x)(x - c)$$

where $q_1(x) = \Sigma a_k(x^{k-1} + cx^{k-2} + \cdots + c^{k-1})$ and

$$(18) \quad f_R(c) = a_0 + a_1c + a_2c^2 + \cdots + a_nc^n.$$

Hence $f(x) = q_1(x)(x - c) + f_R(c)$ and $f_R(c)$ is the right-hand remainder. Similarly, by using the second form of (17) we can prove that the left-hand remainder on division by $x - c$ is

$$(19) \quad f_L(c) = a_0 + ca_1 + c^2a_2 + \cdots + c^na_n.$$

An immediate consequence of these results is

The factor theorem. *The polynomial $(x - c)$ is a right (left) factor of $f(x)$ if and only if c is a right- (left-) hand root in the sense that $f_R(c) = 0$ $(f_L(c) = 0)$.*

If $\mathfrak{A}$ is commutative, we can, of course, drop the modifiers "left" and "right" in the foregoing discussion. If $\mathfrak{A} = \mathfrak{F}$ is a field, the division process can be applied to any pair of polynomials $f(x)$, $g(x) \neq 0$. This fact can be used to prove the important

Theorem 6. *Every ideal in $\mathfrak{F}[x]$, $\mathfrak{F}$ a field, is a principal ideal.*

Proof. Let $\mathfrak{B}$ be an ideal in $\mathfrak{F}[x]$. If $\mathfrak{B} = 0$, the ideal consisting of 0 alone, then $\mathfrak{B} = (0)$, the principal ideal generated by 0. Assume therefore that $\mathfrak{B} \neq 0$. Let $g(x)$ be a non-zero polynomial of least degree in $\mathfrak{B}$. If $f(x)$ is any element of $\mathfrak{B}$, we write $f(x) = g(x)q(x) + r(x)$, where $\deg r(x) < \deg g(x)$. Then $r(x) = f(x) - g(x)q(x) \ \varepsilon \ \mathfrak{B}$, and, since its degree is less than that of $g(x)$, $r(x) = 0$. Hence $f(x) = g(x)q(x)$ is in the principal ideal $(g(x))$. Thus $\mathfrak{B} \subseteq (g(x))$. But $g(x) \ \varepsilon \ \mathfrak{B}$ so that we also have $(g(x)) \subseteq \mathfrak{B}$. Hence $\mathfrak{B} = (g(x))$.

This theorem enables us to state for fields the following sharper form of Theorem 5.

Corollary 1. *If $\mathfrak{F}$ is a field, any polynomial ring $\mathfrak{F}[u] \cong \mathfrak{F}[x]/(g(x))$ where either $g(x) = 0$ or $g(x)$ is a polynomial of positive degree.*

The possibility that $g(x)$ is a non-zero polynomial of 0 degree is excluded since it implies that $(g(x)) = \mathfrak{F}[x]$.

EXERCISES

1. If $f(x) = a_0 + a_1 x + \cdots + a_n x^n$, define $f'(x) = a_1 + 2a_2 x + \cdots + n a_n x^{n-1}$. Prove the usual rules:

$$(f + g)' = f' + g', \quad (cf)' = cf', \quad c \text{ in } \mathfrak{A}$$

$$(fg)' = fg' + f'g.$$

2. Prove *Leibniz's theorem*

$$(fg)^{(k)} = \sum_0^k \binom{k}{i} f^{(i)} g^{(k-i)}$$

where $f^{(i)} = f^{(i-1)'}$, $f^{(0)} = f$.

7. Simple extensions of a field. The methods that we have developed in this chapter can be used to construct field extensions of any given field $\mathfrak{F}$. As we shall see, any such extension can be

obtained by making a succession of *simple* extensions of two types
that we proceed to describe.

Simple transcendental extension. For the given field $\mathfrak{F}$ we construct first the polynomial ring $\mathfrak{F}[x]$, x transcendental. We know
that $\mathfrak{F}[x]$ is an integral domain but not a field. However, we can
imbed $\mathfrak{F}[x]$ in its field of fractions. We denote the latter as $\mathfrak{F}(x)$
and we call its elements *rational expressions* (*functions*) in x over
the base field $\mathfrak{F}$. These elements have the form $f(x)/g(x)$ where
$f(x)$ and $g(x)$ are polynomials and $g(x) \neq 0$. The usual rules of
reckoning hold.

Simple algebraic extensions. This method of extending a field
was used first by Cauchy in defining the field C of complex numbers as an extension of the field R of real numbers. In Cauchy's
case one forms the difference ring $C = R[x]/(x^2 + 1)$ where
$(x^2 + 1)$ is the principal ideal of multiples of $x^2 + 1$. It can
be shown that C is a field extension of R that contains a root
of the equation $x^2 + 1 = 0$. Cauchy's method was generalized
by Kronecker to apply to any field $\mathfrak{F}$ and any polynomial $f(x)$ ε
$\mathfrak{F}[x]$ which is irreducible (prime) in this domain. By saying that
$f(x)$ is *irreducible*, we mean that $f(x)$ cannot be factored as a
product of two polynomials of positive degree. We assume also
that deg $f(x) > 0$.

As in the special case that we have indicated we form the difference ring $\mathfrak{E} = \mathfrak{F}[x]/(f(x))$ where, as usual, $(f(x))$ denotes the
principal ideal generated by $f(x)$. The ring $\mathfrak{E}$ has the identity
$\bar{1} = 1 + (f(x))$ and $\bar{1} \neq 0$ since $f(x)$ is of positive degree. Consider now any coset $\overline{g(x)} = g(x) + (f(x)) \neq 0$. Let $\mathfrak{B}$ be the
totality of polynomials of the form $u(x)g(x) + v(x)f(x)$ where
$u(x)$ and $v(x)$ are arbitrary in $\mathfrak{F}[x]$. It is apparent that $\mathfrak{B}$ is an
ideal in $\mathfrak{F}[x]$. Hence $\mathfrak{B} = (d(x))$. Since $f(x) = 0g(x) + 1f(x)$ ε $\mathfrak{B}$,
$f(x) = d(x)f_1(x)$. Hence either $d(x)$ is a non-zero element of $\mathfrak{F}$
or $d(x)$ is a multiple (by an element of $\mathfrak{F}$) of $f(x)$. On the other
hand, $g(x)$ ε $\mathfrak{B}$ so that $g(x) = d(x)g_1(x)$. Hence if $d(x)$ is a
multiple of $f(x)$, then $g(x)$ is a multiple of $f(x)$ and this contradicts
the assumption that $\overline{g(x)} \neq 0$. Hence we see that $d(x) = d$
is a non-zero element of $\mathfrak{F}$. Since d ε $\mathfrak{B}$, this element has the form
$u(x)g(x) + v(x)f(x)$. If we multiply by d^{-1}, we obtain polynomials $a(x)$, $b(x)$ such that

$$(20) \qquad\qquad a(x)f(x) + b(x)g(x) = 1.$$

The relation (20) gives $\overline{a(x)}\,\overline{f(x)} + \overline{b(x)}\,\overline{g(x)} = \overline{1}$. Since $\overline{f(x)} = 0$, we conclude that $\overline{b(x)}\,\overline{g(x)} = \overline{1}$. Thus any non-zero element of $\mathfrak{E}$ has an inverse. Since $\mathfrak{E}$ is commutative, this means that $\mathfrak{E}$ is a field.

We note next that $\mathfrak{E}$ is an extension of $\mathfrak{F}$. Thus consider the natural homomorphism $g(x) \rightarrow \overline{g(x)}$ of $\mathfrak{F}[x]$ onto $\mathfrak{E}$. This mapping induces a homomorphism of $\mathfrak{F}$ onto a subring $\overline{\mathfrak{F}}$ of $\mathfrak{E}$. The image set $\overline{\mathfrak{F}}$ is the totality of cosets $\bar{a} = a + (f(x))$, a in $\mathfrak{F}$; hence it includes $\overline{1} \neq 0$. On the other hand, $\mathfrak{F}$ is a field. Hence a homomorphic image of it is either 0 or it is isomorphic to $\mathfrak{F}$. It follows that $\overline{\mathfrak{F}} \cong \mathfrak{F}$. In this way $\mathfrak{F}$ is imbedded in $\mathfrak{E}$. As usual we shall identify $\mathfrak{F}$ with $\overline{\mathfrak{F}}$ and write a for the coset $\bar{a}$.

We show finally that $\mathfrak{E} = \mathfrak{F}[\bar{x}]$ and $\bar{x}$ is an algebraic element satisfying the equation $f(\bar{x}) = 0$. First, if $g(x)$ is any polynomial, then $\overline{g(x)} = g(\bar{x})$ is a polynomial in $\bar{x}$ with coefficients in $\mathfrak{F}$. As a matter of fact it is easy to see that any element of $\mathfrak{E}$ can be expressed as a polynomial in $\bar{x}$ of degree $< \deg f(x)$; for we can write $g(x) = f(x)q(x) + r(x)$ where $\deg r(x) < \deg f(x)$. Hence $\overline{g(x)} = \overline{r(x)} = r(\bar{x})$. Since $0 = \overline{f(x)} = f(\bar{x})$, $\bar{x}$ is a root of the equation $f(x) = 0$.

The construction of the difference ring $\mathfrak{E} = \mathfrak{F}[x]/(f(x))$ can also be carried out for reducible polynomials $f(x)$. If $f(x) = f_1(x)f_2(x)$ where $\deg f_i(x) > 0$, then $\overline{f_i(x)} \neq 0$ in $\mathfrak{E}$ but $\overline{f_1(x)f_2(x)} = \overline{f(x)} = 0$. Thus in this case we obtain a ring with zero-divisors $\neq 0$. It is clear at any rate that $\mathfrak{E}$ is commutative and that $\mathfrak{E}$ has an identity.

EXERCISES

1. Let $\mathfrak{E} = R_0[x]/(x^3 + 3x - 2)$. Express the following elements of $\mathfrak{E}$ as polynomials of degree <3 in $\bar{x}$:

(a) $$(2\bar{x}^2 + \bar{x} - 3)(3\bar{x}^2 - 4\bar{x} + 1)$$

(b) $$(2\bar{x}^2 + 4\bar{x} - 5)^{-1}.$$

2. Show that, if $f(x)$ has a square factor $(f(x) = [f_1(x)]^2 f_2(x),\ \deg f_1(x) > 0)$, then $\mathfrak{E} = \mathfrak{F}[x]/(f(x))$ contains non-zero nilpotent elements.

8. Structure of any field. In analyzing the structure of any field $\mathfrak{F}$ we examine first the smallest subfield $\mathfrak{P}$ of $\mathfrak{F}$. We shall call this field *the prime field* of $\mathfrak{F}$. We know that the intersection of any number of subfields of $\mathfrak{F}$ is a subfield. Hence the prime field can be defined to be the intersection of *all* subfields of $\mathfrak{F}$.

We know that $\mathfrak{P}$ contains 1; hence $\mathfrak{P}$ contains the subring $[[1]]$ generated by 1. Now we know that a ring generated by 1 is isomorphic to either I or to $I/(m)$, $m > 0$. (§9 Chapter II). If the second alternative holds here, then $m = p$ is a prime; for otherwise $I/(m)$ has zero-divisors $\neq 0$ and consequently $[[1]]$ has zero-divisors $\neq 0$. But this is clearly impossible in a field. Hence we have the following two possibilities:

I $$[[1]] \cong I$$

II $$[[1]] \cong I/(p), \quad p \text{ a prime.}$$

If I holds, $[[1]]$ is an integral domain but not a field. Hence in order to obtain the prime field we must take the totality of elements of the form $(m1)(n1)^{-1}$ where $m, n \ \varepsilon \ I$ and $n \neq 0$. Thus it is clear that $\mathfrak{P}$ is isomorphic to the field of rational numbers. If II holds, $[[1]]$ is a field since $I/(p)$ is a field. It is clear that in case I $\mathfrak{F}$ has characteristic 0 while in II $\mathfrak{F}$ has characteristic p.

We suppose next that $\mathfrak{F}_0$ is any subfield of $\mathfrak{F}$ and we proceed to determine the structure of the subfield $\mathfrak{F}_0(\theta)$ generated by $\mathfrak{F}_0$ and an additional element θ of $\mathfrak{F}$ (possibly in $\mathfrak{F}_0$). We consider first the subring $\mathfrak{F}_0[\theta]$ generated by $\mathfrak{F}_0$ and θ. We have seen (p. 100) that $\mathfrak{F}_0[\theta] \cong \mathfrak{F}_0[x]/(f(x))$ where either $f(x) = 0$ or $f(x)$ is of positive degree. The ideal $(f(x))$ is the kernel of the homomorphism $g(x) \rightarrow g(\theta)$. Now if $f(x)$ is reducible, then $\mathfrak{E} = \mathfrak{F}_0[x]/(f(x))$ is not an integral domain; hence this possibility is excluded. Thus we have the following two possibilities:

I $$\mathfrak{F}_0[\theta] \cong \mathfrak{F}_0[x]$$

II $$\mathfrak{F}_0[\theta] \cong \mathfrak{F}_0[x]/(f(x)), \quad f(x) \text{ irreducible.}$$

In I, θ is transcendental and $\mathfrak{F}_0(\theta)$ is isomorphic to the field $\mathfrak{F}_0(x)$ of rational expressions in x. In II, $f(\theta) = 0$ so that θ is algebraic. Also in this case $\mathfrak{F}_0[\theta]$ is a field since $\mathfrak{F}_0[x]/(f(x))$ is a field. Hence $\mathfrak{F}_0(\theta) = \mathfrak{F}_0[\theta]$. In either case we see that $\mathfrak{F}_0(\theta)$ is essentially a simple extension of $\mathfrak{F}_0$ of the types considered in the preceding section.

We now know the nature of the prime field of any field and the nature of any subfield $\mathfrak{F}_0(\theta)$. We shall now show that any field can be built up from its prime field by a succession of simple extensions (algebraic or transcendental). A proof of this result for a given field requires that the field be well ordered.* However, the algebraic idea underlying the argument can be fully revealed in considering the countable case. Hence we assume that $\mathfrak{F}$ is countable (finite or denumerably infinite) and we suppose that $\theta_1, \theta_2, \theta_3, \cdots$ is an enumeration of the elements of $\mathfrak{F}$. Set $\mathfrak{F}_0 = \mathfrak{P}$, $\mathfrak{F}_i = \mathfrak{F}_{i-1}(\theta_i)$. Then $\mathfrak{F} = \cup \mathfrak{F}_i$ and each $\mathfrak{F}_i$ is obtained from $\mathfrak{F}_{i-1}$ by a simple transcendental or simple algebraic extension.

9. The number of roots of a polynomial in a field. If $f(x)$ is a polynomial with coefficients in a field and c_1 is a root of $f(x) = 0$, then $f(x) = (x - c_1)f_1(x)$. Suppose now that $c_1, c_2, \cdots, c_m$ are distinct roots of $f(x) = 0$. Then substitution of c_2 in $f(x) = (x - c_1)f_1(x)$ (that is, applying the homomorphism $g(x) \rightarrow g(c_2)$) gives

$$0 = f(c_2) = (c_2 - c_1)f_1(c_2).$$

Since $c_2 \neq c_1, f_1(c_2) = 0$. Hence $f_1(x) = (x - c_2)f_2(x)$ and $f(x) = (x - c_1)(x - c_2)f_2(x)$. Continuing in this way, we can prove that $f(x) = (x - c_1)(x - c_2) \cdots (x - c_m)f_m(x)$. Evidently this implies that the degree n of $f(x) \geq m$. This proves the following

Theorem 7. *If $\mathfrak{F}$ is a field and $f(x)$ is a polynomial of degree $n \geq 0$ with coefficients in $\mathfrak{F}$, then $f(x)$ has at most n distinct roots in $\mathfrak{F}$.*

EXERCISES

1. If $a_n \not\equiv 0 \pmod{p}$, then the congruence $a_0 + a_1x + \cdots + a_nx^n \equiv 0 \pmod{p}$ has at most n incongruent solutions in I.

2. Prove that, if $\mathfrak{F}$ is a finite field containing q elements a_i, then $h(x) = x^q - x = (x - a_1)(x - a_2) \cdots (x - a_q)$ in $\mathfrak{F}[x]$.

3. Prove that, if p is a prime integer, then $(p - 1)! \equiv -1 \pmod{p}$. This is known as *Wilson's theorem*.

4. Show that the polynomial $x^3 - x$ has 6 roots in $I/(6)$.

5. Show that the polynomial $x^2 + 1$ has an infinite number of roots in the ring Q of real quaternions.

* For a discussion of well ordering, consult van der Waerden's *Moderne Algebra*, vol. 1, 1st ed., chapter 8.

10. Polynomials in several elements. Again let $\mathfrak{B}$ be a ring with an identity and let $\mathfrak{A}$ be any subring containing 1. Suppose that $u_1, u_2, \cdots, u_r$ are elements of $\mathfrak{B}$ that commute with each other and that commute with every $a \in \mathfrak{A}$. Let $\mathfrak{A}[u_1, u_2, \cdots, u_r]$ denote the subring generated by $\mathfrak{A}$ and by the u_i and write $\mathfrak{A}[u_1][u_2] \cdots [u_r]$ for $(((\mathfrak{A}[u_1])[u_2]) \cdots)[u_r]$. We assert that

$$(21) \qquad \mathfrak{A}[u_1, u_2, \cdots, u_r] = \mathfrak{A}[u_1][u_2] \cdots [u_r].$$

This is clear for $r = 1$. Hence we assume it for $s - 1$ and we consider $\mathfrak{A}[u_1, u_2, \cdots, u_s]$. This ring contains $\mathfrak{A}[u_1, \cdots, u_{s-1}]$ and the element u_s. Hence it contains $\mathfrak{A}[u_1, \cdots, u_{s-1}][u_s]$. On the other hand, $\mathfrak{A}[u_1, \cdots, u_{s-1}][u_s]$ is a subring that contains $u_1, u_2, \cdots, u_s$. Hence it contains $\mathfrak{A}[u_1, \cdots, u_s]$. Thus we have

$$\mathfrak{A}[u_1, \cdots, u_s] = \mathfrak{A}[u_1, \cdots, u_{s-1}][u_s]$$
$$= \mathfrak{A}[u_1] \cdots [u_{s-1}][u_s]$$

by the induction assumption.

By (21), or directly, we can see that $\mathfrak{A}[u_1, u_2, \cdots, u_r]$ is the totality of *polynomials*

$$\Sigma a_{i_1 i_2 \cdots i_r} u_1^{i_1} u_2^{i_2} \cdots u_r^{i_r}$$

in the u's with coefficients $a_{i_1 i_2 \cdots i_r}$ in $\mathfrak{A}$. As a generalization of the notion of transcendental element we now define the elements $u_1, u_2, \cdots, u_r$ to be *algebraically independent over* $\mathfrak{A}$ if the only relation of the form

$$(22) \qquad \Sigma d_{i_1 i_2 \cdots i_r} u_1^{i_1} u_2^{i_2} \cdots u_r^{i_r} = 0,$$

$d_{i_1 i_2 \cdots i_r}$ in $\mathfrak{A}$, that holds for the u's is that in which all the d's are 0. Since the u's commute, it is clear that this condition does not depend on the order of the elements $u_1, u_2, \cdots, u_r$. Moreover, it is clear that, according to the definition, u_1 is algebraically independent over $\mathfrak{A}$ if and only if it is transcendental. We now prove the following more general result.

Lemma. *The elements* $u_1, u_2, \cdots, u_r$ *are algebraically independent over* $\mathfrak{A}$ *if and only if each* u_k, $k = 1, 2, \cdots, r$, *is transcendental over* $\mathfrak{A}[u_1, u_2, \cdots, u_{k-1}]$.

Proof. Suppose that each u_k, $k = 1, 2, \cdots, r$, is transcendental over $\mathfrak{A}[u_1, \cdots, u_{k-1}]$ and assume that (22) holds. Write this relation as

$$(23) \qquad D_0 + D_1 u_r + D_2 u_r^2 + \cdots + D_m u_r^m = 0$$

where $D_i = \Sigma d_{i_1 i_2 \cdots i_{r-1} i} u_1^{i_1} u_2^{i_2} \cdots u_{r-1}^{i_{r-1}}$. Then each $D_i = 0$ and, using induction, we can assume that this implies $d_{i_1 i_2 \cdots i_{r-1} i} = 0$ for all i_1, i_2, $\cdots$. Hence the u_i are algebraically independent. Conversely, suppose that u_1, u_2, $\cdots$, u_r form an algebraically independent set, and assume that we have a relation of the form $\Sigma D_i u_k^i = 0$ where the $D_i \ \varepsilon \ \mathfrak{A}[u_1, u_2, \cdots, u_{k-1}]$. We can write $D_i = \Sigma d_{i_1 i_2 \cdots i_{k-1} i} u_1^{i_1} u_2^{i_2} \cdots u_{k-1}^{i_{k-1}}$ and obtain $\Sigma d_{i_1 \cdots i_{k-1} i} u_1^{i_1} u_2^{i_2} \cdots u_{k-1}^{i_{k-1}} u_k^{i} = 0$. Then $d_{i_1 i_2 \cdots i_{k-1} i} = 0$ for all i_1, i_2, $\cdots$, i and $D_i = 0$ for all i. Hence u_k is transcendental over $\mathfrak{A}[u_1, u_2, \cdots, u_{k-1}]$.

This lemma enables us to construct inductively for any given ring $\mathfrak{A}$ with an identity a ring $\mathfrak{B} = \mathfrak{A}[x_1, x_2, \cdots, x_r]$ where the x_i are algebraically independent over $\mathfrak{A}$; for we can construct successively the rings $\mathfrak{A}[x_1]$, $\mathfrak{A}[x_1][x_2]$, $\cdots$ in which each x_k is transcendental over $\mathfrak{A}[x_1] \cdots [x_{k-1}] = \mathfrak{A}[x_1, \cdots, x_{k-1}]$. Then it is clear that $\mathfrak{A}[x_1] \cdots [x_r] = \mathfrak{A}[x_1, \cdots, x_r]$ is a ring of the required type.

If the x_i are algebraically independent over $\mathfrak{A}$ and the y_i, $i = 1, 2, \cdots, r$ are algebraically independent over $\mathfrak{A}$, then $\mathfrak{A}[x_1, x_2, \cdots, x_r]$ is isomorphic to $\mathfrak{A}[y_1, y_2, \cdots, y_r]$. This is an immediate consequence of the following theorem.

Theorem 8. *Let $\mathfrak{A}_i$, $i = 1, 2$, be a ring with an identity and let $\mathfrak{A}_i[x_{1i}, x_{2i}, \cdots, x_{ri}]$ be a ring of polynomials in the algebraically independent elements x_{ji}. Then any homomorphism (isomorphism) of $\mathfrak{A}_1$ onto $\mathfrak{A}_2$ can be extended in one and only one way to a homomorphism (isomorphism) of $\mathfrak{A}_1[x_{11}, x_{21}, \cdots, x_{r1}]$ onto $\mathfrak{A}_2[x_{12}, x_{22}, \cdots, x_{r2}]$ mapping x_{j1} into x_{j2} for $j = 1, 2, \cdots, r$.*

The case $r = 1$ of this theorem has been proved in the preceding section. The extension to arbitrary r is immediate by induction. The details of the argument will be left to the reader.

The same inductive procedure also yields the following two results: (1) If $\mathfrak{A}$ is an integral domain, then so is $\mathfrak{A}[x_1, x_2, \cdots, x_r]$.

(2) If $\mathfrak{A}$ is an integral domain, then the only units of $\mathfrak{A}[x_1, x_2, \cdots, x_r]$ are the elements of $\mathfrak{A}$ that are units in $\mathfrak{A}$.

EXERCISE

1. Show that a ring $\mathfrak{A}[x_1, x_2, \cdots, x_r]$, x_i algebraically independent, can also be obtained as a semi-group ring over $\mathfrak{A}$ of the semi-group S of r-tuples $(i_1, i_2, \cdots, i_r)$ of non-negative integers i_j where the composition is

$$(i_1, i_2, \cdots, i_r)(j_1, j_2, \cdots, j_r) = (i_1 + j_1, i_2 + j_2, \cdots, i_r + j_r).$$

***11. Symmetric polynomials.**   Suppose that the elements x_i of $\mathfrak{A}[x_1, x_2, \cdots, x_r]$ are algebraically independent.   Clearly if $x_{1'}, x_{2'}, \cdots, x_{r'}$ is any permutation of $x_1, x_2, \cdots, x_r$, then $\mathfrak{A}[x_1, x_2, \cdots, x_r] = \mathfrak{A}[x_{1'}, x_{2'}, \cdots, x_{r'}]$.   Hence we can conclude from the preceding theorem that the mapping

$$(24) \qquad \Sigma a_{i_1 i_2 \cdots i_r} x_1{}^{i_1} x_2{}^{i_2} \cdots x_r{}^{i_r} \to \Sigma a_{i_1 i_2 \cdots i_r} x_{1'}{}^{i_1} x_{2'}{}^{i_2} \cdots x_{r'}{}^{i_r}$$

is an automorphism of $\mathfrak{A}[x_1, x_2, \cdots, x_r]$.   Thus the permutation

$$\sigma : \begin{pmatrix} x_1 & x_2 & \cdots & x_r \\ x_{1'} & x_{2'} & \cdots & x_{r'} \end{pmatrix}$$ of the x's can be extended in one and only one way to an automorphism σ^* of $\mathfrak{A}[x_1, x_2, \cdots, x_r]$ that acts as the identity in $\mathfrak{A}$.

Now if A and B are automorphisms of a ring, then the resultant AB is also an automorphism.   In particular, if σ^* and τ^* are the automorphisms determined by the elements σ, τ of S_r, then $\sigma^* \tau^*$ is an automorphism of $\mathfrak{A}[x_1, x_2, \cdots, x_r]$.   Now the automorphisms $\sigma^* \tau^*$ and $(\sigma\tau)^*$ effect the same permutation $\sigma\tau$ on the x_i and effect the identity mapping in the coefficient ring $\mathfrak{A}$.   From this it follows that $\sigma^* \tau^* = (\sigma\tau)^*$.   Hence the set Σ of the automorphisms σ^* is a transformation group isomorphic to the symmetric group S_r.

A polynomial $f(x_1, x_2, \cdots, x_r)$ is said to be *symmetric* in the x's if $f\sigma^* = f$ for all $\sigma^* \, \varepsilon \, \Sigma$.   The totality of these polynomials constitute a subring $\mathfrak{S}$ of $\mathfrak{A}[x_1, x_2, \cdots, x_r]$.   Evidently $\mathfrak{S} \supseteq \mathfrak{A}$.   Also the coefficients of the polynomial

$$F(x) = (x - x_1)(x - x_2) \cdots (x - x_r)$$

are symmetric; for we can extend the automorphism σ^* of $\mathfrak{A}[x_1, x_2, \cdots, x_r]$ to an automorphism σ^{**} of $\mathfrak{A}[x_1, \cdots, x_r; x]$ so that $x\sigma^{**} = x$.   The extension σ^{**} permutes the factors of $F(x)$ and therefore it maps $F(x)$ into itself. It follows that the coefficients of $F(x)$ are left unchanged by σ^{**} and consequently by σ^{*}. Since this holds for all σ, the coefficients of $F(x)$ are symmetric. We can calculate these coefficients and see that

$$F(x) = x^r - p_1 x^{r-1} + p_2 x^{r-2} - \cdots + (-1)^r p_r$$

where

$$(25) \qquad p_1 = \sum_i x_i, \quad p_2 = \sum_{i<j} x_i x_j, \quad p_3 = \sum_{i<j<k} x_i x_j x_k, \cdots,$$

$$p_r = x_1 x_2 \cdots x_r.$$

We shall call the p_i *elementary symmetric polynomials*, and we shall prove that $\mathfrak{S} = \mathfrak{A}[p_1, p_2, \cdots, p_r]$ and that the p_i are algebraically independent over $\mathfrak{A}$.

The equation $\mathfrak{S} = \mathfrak{A}[p_1, p_2, \cdots, p_r]$ means, of course, that every symmetric polynomial can be expressed as a polynomial in the elementary symmetric functions p_i. It suffices to prove this for homogeneous polynomials. By a *homogeneous polynomial* we mean one in which all of the terms $a x_1^{k_1} x_2^{k_2} \cdots x_r^{k_r}$ have the same *total degree* $k = k_1 + k_2 + \cdots + k_r$. Any polynomial can be expressed in one and only one way as a sum of homogeneous polynomials of different degrees. Since the automorphisms σ^{*} preserve degree, it is clear that, if $f(x_1, x_2, \cdots, x_r)$ is symmetric, then so are its homogeneous parts.

We suppose now that $f(x_1, x_2, \cdots, x_r)$ is a homogeneous symmetric polynomial of degree, say m. We shall introduce the lexicographic ordering for the monomials of degree m, that is, we say that $a x_1^{k_1} x_2^{k_2} \cdots x_r^{k_r}$ is *higher* than $b x_1^{l_1} x_2^{l_2} \cdots x_r^{l_r}$ if $k_1 = l_1, k_2 = l_2, \cdots, k_s = l_s$ but $k_{s+1} > l_{s+1} (s \geq 0)$. Thus, for example, $x_1^2 x_2 x_3 > x_1 x_2^3 > x_1 x_2^2 x_3$. Now let $a x_1^{k_1} x_2^{k_2} \cdots x_r^{k_r}$ be the highest term in f. Then since f contains all the terms that can be obtained from $a x_1^{k_1} x_2^{k_2} \cdots x_r^{k_r}$ by permuting the x's, it is clear that $k_1 \geq k_2 \geq k_3 \geq \cdots \geq k_r$ in the highest term of f.

We consider now the highest term of the homogeneous symmetric polynomial $p_1^{d_1} p_2^{d_2} \cdots p_r^{d_r}$. Using the definitions (25) we can see that this term is

$$x_1^{d_1+d_2+\cdots+d_r} x_2^{d_2+\cdots+d_r} \cdots x_r^{d_r}.$$

Hence the highest term of $ap_1^{k_1-k_2}p_2^{k_2-k_3} \cdots p_r^{k_r}$ is the same as that of f and hence the highest term of the homogeneous symmetric polynomial $f_1 = f - ap_1^{k_1-k_2}p_2^{k_2-k_3} \cdots p_r^{k_r}$ is less than that of f. We can repeat our process with f_1. Since there are only a finite number of highest terms that are lower than a given one, a finite number of applications of this process yields a representation of f as a polynomial in the p_i.

We shall show now that the elementary symmetric polynomials are algebraically independent. If any of the coefficients in our relation are $\neq 0$, we consider the set of exponents $(d_1, d_2, \cdots, d_r)$ for which $a_{d_1 \cdots d_r} \neq 0$. Introduce

$$k_1 = d_1 + d_2 + \cdots + d_r, \quad k_2 = d_2 + \cdots + d_r, \quad \cdots, \quad k_r = d_r.$$

Then the highest term in the lexicographic ordering in $a_{d_1 \cdots d_r}p_1^{d_1}p_2^{d_2} \cdots p_r^{d_r}$ is $a_{d_1 \cdots d_r}x_1^{k_1}x_2^{k_2} \cdots x_r^{k_r}$. If $(d_1', d_2', \cdots, d_r')$ is a second set of exponents such that $a_{d_1' \cdots d_r'} \neq 0$, then $a_{d_1' \cdots d_r'}p_1^{d_1'}p_2^{d_2'} \cdots p_r^{d_r'}$ has as its highest term $a_{d_1' \cdots d_r'}x_1^{k_1'}x_2^{k_2'} \cdots x_r^{k_r'}$ where $k_i' = d_i' + d_{i+1}' + \cdots + d_r'$, $i = 1, 2, \cdots, r$. Clearly if $k_i = k_i'$ then $d_i = d_i'$ for all i. Thus distinct terms in the p's have distinct highest terms in the x's. If we choose the term $a_{d_1 \cdots d_r}p_1^{d_1}p_2^{d_2} \cdots p_r^{d_r}$ so that $x_1^{k_1}x_2^{k_2} \cdots x_r^{k_r}$ is higher than any other $x_1^{k_1'}x_2^{k_2'} \cdots x_r^{k_r'}$, it is clear that the term $x_1^{k_1}x_2^{k_2} \cdots x_r^{k_r}$ occurs only once in the relation for the p's. This gives a nontrivial relation for the x's and contradicts the algebraic independence of the latter elements. This proves the second part of the following

Theorem 9. *Every symmetric polynomial is expressible as a polynomial in the elementary symmetric polynomials p_i defined in* (25). *The elementary symmetric polynomials $p_1, p_2, \cdots, p_r$ are algebraically independent over* $\mathfrak{A}$. *Every x_i is algebraic over* $\mathfrak{A}[p_1, p_2, \cdots, p_r]$.

The last statement of the theorem is clear since

$$F(x_i) = x_i^r - p_1x_i^{r-1} + \cdots + (-1)^r p_r = 0.$$

EXERCISES

1. Express $\displaystyle\sum_{i,j,k,\neq} x_i^2 x_j^2 x_k$ $(n \geq 5)$ in terms of the elementary symmetric functions.

2. Let $\Delta = \displaystyle\prod_{i<j} (x_i - x_j)$. Show that if η is a transposition then $\Delta\eta^* = -\Delta$. Use this to prove that if τ is a permutation that has a decomposition as a product of an even (odd) number of transpositions then any factorization of τ as a product of transpositions contains an even (odd) number of terms.

3. Show that Δ^2 is symmetric. Express Δ^2 for $r = 3$ in terms of the elementary symmetric functions.

4. Show that the symmetric polynomials $s_k = \Sigma x_i^k$ satisfy *Newton's identities*.

$$s_k - p_1 s_{k-1} + p_2 s_{k-2} - \cdots + (-1)^{k-1} p_{k-1} s_1 + (-1)^k k p_k = 0, \quad k = 1, 2, \cdots, n.$$

12. Rings of functions. Let S be an arbitrary non-vacuous set and let $\mathfrak{A}$ be an arbitrary ring. Consider the totality $(\mathfrak{A},S)$ of functions with domain S and with range contained in $\mathfrak{A}$. Thus the elements f of $(\mathfrak{A},S)$ are the mappings $s \to f(s)$ of S into $\mathfrak{A}$. (Note that the effect of f on s is denoted here in the conventional manner as $f(s)$ rather than as sf as is usual in these *Lectures*.) As usual $f = g$ means that $f(s) = g(s)$ for all $s \in S$. Now we define addition and multiplication in $(\mathfrak{A},S)$ in the customary way by

$$(26) \qquad\qquad (f + g)(s) = f(s) + g(s)$$

$$(fg)(s) = f(s)g(s).$$

It can be easily verified that $(\mathfrak{A},S)$ with these compositions is a ring; for the associativity of addition and multiplication, the commutativity of addition and the distributive laws follow immediately from the corresponding laws in $\mathfrak{A}$. For example, we have

$$((f + g)h)(s) = (f(s) + g(s))h(s) = f(s)h(s) + g(s)h(s)$$

$$= (fh + gh)(s).$$

Hence $(f + g)h = fh + gh$. The function 0 such that $0(s) = 0$ for all s acts as the identity under addition and $-f$ is the function such that $(-f)(s) = -f(s)$ for all s.

If a is any element of $\mathfrak{A}$, we define *the constant function a* by the requirement that $a(s) = a$ for all s. These functions constitute a subring of $(\mathfrak{A},S)$ isomorphic to $\mathfrak{A}$. We denote this subring by $\mathfrak{A}$ also. If $\mathfrak{A}$ has an identity, then the associated constant function acts as the identity in the whole ring $(\mathfrak{A},S)$.

For the sake of simplicity we shall now assume that $\mathfrak{A}$ is a commutative ring with an identity. We consider the ring of functions $\bar{\mathfrak{A}} = (\mathfrak{A},\mathfrak{A})$. In addition to the constant functions a particularly important function is the identity function $s \rightarrow s$. We use the customary notation s for this function as for the variable s in $\mathfrak{A}$. Since $\mathfrak{A}$ is commutative, this function commutes with the constant functions. We call the elements of the ring $\mathfrak{A}[s]$ generated by the constant functions and by the identity function *polynomial functions in one variable*. If $f(x) = a_0 + a_1 x + \cdots + a_n x^n$ is an element of $\mathfrak{A}[x]$ where x is transcendental, then $f(s)$ is the function that maps s into the element $a_0 + a_1 s + \cdots + a_n s^n$ of $\mathfrak{A}$, and $\mathfrak{A}[s]$ is the totality of these functions.

The function s need not be transcendental over $\mathfrak{A}$. Thus if $\mathfrak{A}$ is a finite ring with elements $a_1, a_2, \cdots, a_q$, then the polynomial

$$(27) \qquad h(x) = (x - a_1)(x - a_2) \cdots (x - a_q) \neq 0,$$

while the function

$$(28) \qquad h(s) = (s - a_1)(s - a_2) \cdots (s - a_q) = 0.$$

This is clear since the element $h(s) = 0$ for all $s \, \varepsilon \, \mathfrak{A}$. If $\mathfrak{A}$ is a finite field, then we know that $h(x) = x^q - x$ (ex. 2, p. 104).

On the other hand, we shall now show that, if $\mathfrak{A} = \mathfrak{F}$ is an infinite field, then the identity function is transcendental. This is an immediate consequence of Theorem 7 (§ 9); for, if $f(x)$ is a polynomial $\neq 0$ in $\mathfrak{F}[x]$, then $f(s) = 0$ for only a finite number of elements of $\mathfrak{F}$. Hence there exist elements $c \, \varepsilon \, \mathfrak{F}$ such that $f(c) \neq 0$. This means that the function $f(s) \neq 0$ and that s is transcendental.

The definition of polynomial functions in several variables is an immediate generalization of the foregoing. Here we begin with the set $S = \mathfrak{A}^{(r)}$ of r-tuples $(s_1, s_2, \cdots, s_r)$, s_i in $\mathfrak{A}$ and we consider the ring of functions $\bar{\mathfrak{A}}^{(r)} = (\mathfrak{A},\mathfrak{A}^{(r)})$. In this ring we select the particular functions s_i defined by

$$(29) \qquad (s_1, s_2, \cdots, s_r) \rightarrow s_i.$$

Then we define *polynomial functions in r variables* to be elements of the ring $\mathfrak{A}[s_1, s_2, \cdots, s_r]$ generated by the constant functions and by the r functions s_i. Clearly the s_i commute and commute with the constant functions.

If $f(x_1, x_2, \cdots, x_r) \; \varepsilon \; \mathfrak{A}[x_1, x_2, \cdots, x_r]$ where the x_i are algebraically independent, then it is clear what is meant by the function $f(s_1, s_2, \cdots, s_r)$. This function is a polynomial function and every polynomial function is obtained in this way.

If $\mathfrak{A}$ is a finite ring of q elements a_j, then

$$h(s_i) = (s_i - a_1)(s_i - a_2) \cdots (s_i - a_q) = 0.$$

Thus the functions $s_1, s_2, \cdots, s_r$ are algebraic relative to the subring of constant functions. In contrast to this result we shall prove that, if $\mathfrak{F}$ is an infinite field, then the functions s_i are algebraically independent. This result is equivalent to the following

Theorem 10. *If $\mathfrak{F}$ is an infinite field and $f(x_1, x_2, \cdots, x_r)$ is a polynomial $\neq 0$ in the polynomial domain $\mathfrak{F}[x_1, x_2, \cdots, x_r]$, x_i algebraically independent, then there exist elements $c_1, c_2, \cdots, c_r$, in $\mathfrak{F}$ such that $f(c_1, c_2, \cdots, c_r) \neq 0$.*

Proof. The case $r = 1$ has been proved above. Hence we assume that the theorem holds for $r - 1$ x's. We write

$$f(x_1, x_2, \cdots, x_r) = B_0 + B_1 x_r + B_2 x_r{}^2 + \cdots + B_n x_r{}^n$$

where $B_i \; \varepsilon \; \mathfrak{F}[x_1, x_2, \cdots, x_{r-1}]$. Also we can suppose that $B_n \equiv B_n(x_1, x_2, \cdots, x_{r-1}) \neq 0$. Then by the induction assumption we know that there exist elements c_i in $\mathfrak{F}$ such that $B_n(c_1, c_2, \cdots, c_{r-1}) \neq 0$. Thus

$$f(c_1, c_2, \cdots, c_{r-1}, x_r) = B_0(c_1, c_2, \cdots, c_{r-1})$$
$$+ \; B_1(c_1, c_2, \cdots, c_{r-1})x_r + \cdots$$
$$+ \; B_n(c_1, c_2, \cdots, c_{r-1})x_r{}^n \neq 0.$$

Hence we can choose a value $x_r = c_r$ such that $f(c_1, c_2, \cdots, c_r) \neq 0$.

EXERCISES

1. Prove the following extension of the foregoing theorem: If $f(x_1, x_2, \cdots, x_r)$ is a polynomial with coefficients in an infinite field $\mathfrak{F}$ such that $f(c_1, c_2, \cdots, c_r) = 0$ for all $(c_1, c_2, \cdots, c_r)$ for which a second polynomial $g(x_1, x_2, \cdots, x_r) \neq 0$ has values $g(c_1, c_2, \cdots, c_r) \neq 0$, then $f(x_1, x_2, \cdots, x_r) = 0$.

2. Let $\mathfrak{F}$ be a finite field containing q elements. Prove that, if $f(x_1, x_2, \cdots, x_r)$ is a non-zero polynomial of degree $< q$ in each x_i, then there exist c_i in $\mathfrak{F}$ such that $f(c_1, c_2, \cdots, c_r) \neq 0$.

In the remainder of these exercises $\mathfrak{F}$ is as in 2.

3. Prove that every function in r variables (element of $\widetilde{\mathfrak{F}}^{(r)}$) is a polynomial function. (Hint: enumerate the set of functions and the set of polynomial functions.)

4. Show that any polynomial in $\mathfrak{F}[x_1, x_2, \cdots, x_r]$ can be written in the form

$$\sum_{i=1}^{r} g_i(x_1, x_2, \cdots, x_r)(x_i{}^q - x_i) + g_0(x_1, x_2, \cdots, x_r)$$ where g_0 is of degree $< q$ in each x_i.

5. Prove that, if $m(x_1, x_2, \cdots, x_r)$ is a polynomial such that the function $m(s_1, s_2, \cdots, s_r) = 0$, then $m(x_1, x_2, \cdots, x_r)$ can be written in the form $\Sigma g_i(x_1, x_2, \cdots, x_r)(x_i{}^q - x_i)$.

6. Let $f(x_1, x_2, \cdots, x_r)$ be a polynomial such that $f(0, 0, \cdots, 0) = 0$ and $f(c_1, c_2, \cdots, c_r) \neq 0$ for all $(c_1, c_2, \cdots, c_r) \neq (0, 0, \cdots, 0)$. Prove that, if $F(x_1, x_2, \cdots, x_r) = 1 - f(x_1, x_2, \cdots, x_r)^{q-1}$, then

$$F(c_1, c_2, \cdots, c_r) = \begin{cases} 1 \text{ if } (c_1, c_2, \cdots, c_r) = (0, 0, \cdots, 0) \\ 0 \text{ otherwise} \end{cases}.$$

7. Show that the F of 6 determines the same function as

$$F_0 = (1 - x_1{}^{q-1})(1 - x_2{}^{q-1}) \cdots (1 - x_r{}^{q-1}).$$

Hence prove that $\deg F \geq r(q - 1)$ ($\deg F =$ total degree of F).

8. (Artin-Chevalley.) Let $f(x_1, x_2, \cdots, x_r)$ be a polynomial of degree $n < r$ and suppose that $f(0, 0, \cdots, 0) = 0$. Show that there exists a $(c_1, c_2, \cdots, c_r) \neq (0, 0, \cdots, 0)$ such that $f(c_1, c_2, \cdots, c_r) = 0$.

Chapter *IV*

ELEMENTARY FACTORIZATION THEORY

In this chapter we consider the problem of decomposing elements of a given commutative integral domain as products of irreducible elements. In a number of important integral domains such factorizations exist for all the non-units, and in a certain sense uniqueness of factorization holds. In these instances we can determine all of the factors of a given element and hence we can give simple conditions for the solvability of equations of the form $ax = b$. Since the factorization theory that we shall consider is a purely multiplicative theory that concerns the semi-group of non-zero elements of a commutative integral domain, we shall find it clearer to begin our discussion with the factorization theory of semi-groups.

1. Factors, associates, irreducible elements. Let $\mathfrak{S}$ be an arbitrary commutative semi-group that has an identity 1 and that satisfies the cancellation law. If $\mathfrak{U}$ denotes the set of units of $\mathfrak{S}$, then we know that $\mathfrak{U}$ is a subgroup of $\mathfrak{S}$.

If a and b are elements of $\mathfrak{S}$, we say that b is a *factor* or *divisor* of a if there exists an element c in $\mathfrak{S}$ such that $a = bc$. If b is a factor of a, we write $b \mid a$. It is immediate that this relation is transitive and reflexive. An element u is a unit if and only if $u \mid 1$. The units are the trivial factors since they are factors of every element of $\mathfrak{S}$. If $a \mid b$ and $b \mid a$, then we shall say that these elements are *associates*. The conditions for this relation are that $b = au$, $a = bv$. Hence $b = au = bvu$. By the cancellation law $vu = 1$. Thus a and b differ by unit factors. The converse is immediate also and it is clear that the relation of associateness is an equivalence. If a and b are associates, then we write $a \sim b$.

If $b \mid a$ and b is neither a unit nor an associate of a, then we say that b is a *proper factor* of a. In this case $a = bc$ and c is neither a unit nor an associate. Hence c, too, is a proper factor of a. If u is a unit and $u = vw$, then it is immediate that v and w are units. Thus the units of $\mathfrak{S}$ do not have proper factors.

An element a is said to be *irreducible* if a is not a unit and a has no proper factors in $\mathfrak{S}$.

2. Gaussian semi-groups. If an element a of a commutative semi-group $\mathfrak{S}$ has a factorization $a = p_1 p_2 \cdots p_s$ where the p_i are irreducible, then a also has the factorization $a = p_1' p_2' \cdots p_s'$ where $p_i' = u_i p_i$ and the u_i are units such that $u_1 u_2 \cdots u_s = 1$. It is clear that the p_i' are irreducible. Hence if $\mathfrak{S}$ has units $\neq 1$ and $s > 1$, then we can always alter a factorization in the way indicated to obtain other factorizations of the given element. The new factorizations will be regarded as essentially the same as the original one, and we shall say that a factorization $a = p_1 p_2 \cdots p_s$ of a into irreducible elements is *essentially unique* if for any other factorization $a = p_1' p_2' \cdots p_t'$, p_i' irreducible, we have $t = s$ and $p_i' \sim p_i$ for a suitable arrangement of the p_i'. We use this concept to formulate the following

Definition 1. *A semi-group $\mathfrak{S}$ is called* Gaussian *if (1) $\mathfrak{S}$ is commutative, has an identity and satisfies the cancellation law, and (2) every non-unit of $\mathfrak{S}$ has an essentially unique factorization into irreducible elements. An integral domain is* Gaussian *if its semi-group of non-zero elements is Gaussian.*

Our main purpose in this chapter is to show that a number of important types of integral domains are Gaussian. That this is not a universal property can be seen by considering the following

Example. Let $\mathfrak{A} = I[\sqrt{-5}\,]$, the set of complex numbers of the form $a + b\sqrt{-5}$ where a and b are integers. It is easy to see that $\mathfrak{A}$ is a subring of the field of complex numbers. Hence $\mathfrak{A}$ is a commutative integral domain. Also $\mathfrak{A}$ has the identity $1 = 1 + 0\sqrt{-5}$.

The investigation of the arithmetic of $\mathfrak{A}$ is greatly facilitated by the introduction of the norm of elements of this domain. If $r = a + b\sqrt{-5}$, we define the norm $N(r) = r\bar{r} = a^2 + 5b^2$. This function is multiplicative: $N(rs) = N(r)N(s)$ and its values are positive integers for the non-zero elements of $\mathfrak{A}$.

We use the norm first to determine the units of $\mathfrak{A}$. If $rs = 1$, then $N(r)N(s) = N(1) = 1$. Hence $N(r) = a^2 + 5b^2 = 1$. Hence $a = \pm 1$ and $b = 0$. Thus $r = \pm 1$.

It follows that the only associates of an element in $\mathfrak{A}$ are the element and its negative.

We consider now the two factorizations

$$9 = 3 \cdot 3 = (2 + \sqrt{-5})(2 - \sqrt{-5}).$$

Each of the factors, 3 and $2 \pm \sqrt{-5}$, is irreducible. For suppose that $3 = rs$. Then $9 = N(3) = N(r)N(s)$. Hence $N(r) = 1$, 3, or 9. But if $N(r) = 3$, $a^2 + 5b^2 = 3$, and this is impossible for integers a and b. Hence either $N(r) = 1$ or $N(r) = 9$ and $N(s) = 1$. In the first case r is a unit and in the second s is a unit. In a similar manner we see that $2 \pm \sqrt{-5}$ is irreducible. Hence the displayed factorizations are essentially distinct factorizations into irreducible elements and $\mathfrak{A}$ is not Gaussian.

In any Gaussian semi-group $\mathfrak{S}$ one can determine to within unit factors all the factors of a given non-unit element a, provided that a factorization of a into irreducible elements is known; for if $a = p_1 p_2 \cdots p_s$ where the p_i are irreducible, and if $a = bc$ where $b = p_1' p_2' \cdots p_t'$, $c = p_1'' p_2'' \cdots p_u''$ and the p_j' and p_k'' are irreducible, then

$$a = p_1 p_2 \cdots p_s = p_1' p_2' \cdots p_t' p_1'' p_2'' \cdots p_u''.$$

Hence by the uniqueness property $p_j' \sim p_{i_j}$ where $i_j \neq i_k$ if $j \neq k$. Hence $b \sim p_{i_1} p_{i_2} \cdots p_{i_t}$. Thus any factor of a is an associate of one of the 2^s products obtained in this way. If we call the number s of irreducible factors of a the *length* of this element, we see also that any proper factor of a has a smaller length than a. Hence it is clear that any Gaussian semi-group satisfies the following condition:

A. $\mathfrak{S}$ contains no infinite sequences $a_1, a_2, \cdots$ with the property that each a_{i+1} is a proper factor of a_i.

We shall now show that this condition and a second condition that involves the concept of a prime element are sufficient that a commutative semi-group with identity and cancellation law be Gaussian. An element p of $\mathfrak{S}$ is called a *prime* if for any product ab that is divisible by p it is true that either a or b is divisible by p. Our second condition now reads as follows:

B. Every irreducible element of $\mathfrak{S}$ is prime.

Condition A guarantees the existence of a factorization into irreducible elements for any non-unit in $\mathfrak{S}$. Let a be a non-unit. We shall show first that a has an irreducible factor. If a is irreducible, there is nothing to prove. Otherwise let $a = a_1 b_1$ where a_1 is a proper factor. Either a_1 is irreducible or $a_1 = a_2 b_2$ where a_2 is a proper factor of a_1. We continue this process and obtain a sequence $a, a_1, a_2, \cdots$ where each a_i is a proper factor of a_{i-1}. After a finite number of steps this process breaks off by A. If a_n is the last term, a_n is irreducible and $a_n \mid a$.

We now set $a_n = p_1$ and we write $a = p_1 a'$. If a' is a unit, a is irreducible. Otherwise we have $a' = p_2 a''$ where p_2 is irreducible. Continuing in this way, we obtain the sequence $a, a', a'', \cdots$ each a proper factor of the preceding and each $a^{(i-1)} = p_i a^{(i)}$, p_i irreducible. This breaks off with an irreducible element $a^{(s-1)} = p_s$. Then

$$a = p_1 a' = p_1 p_2 a'' = \cdots = p_1 p_2 \cdots p_s$$

where the p_i are irreducible.

We shall show next that condition B insures uniqueness of factorization into irreducible elements; for let

$$(1) \qquad a = p_1 p_2 \cdots p_s = p_1' p_2' \cdots p_t'$$

be two factorizations of an element into irreducible elements. We suppose also that any element that has a factorization as a product of $s - 1$ irreducible elements has essentially only one such factorization. Now the element p_1 in (1) is irreducible; hence, by B, it is prime. A simple inductive argument shows that, if a product of more than two factors is divisible by p_1, then so is one of its factors. This implies that one of the p_i' is divisible by p_1. By rearranging the p' if necessary, we may suppose that p_1' is divisible by p_1. Since p_1 and p_1' are irreducible, this means that $p_1' \sim p_1$ so that $p_1' = p_1 u_1$, u_1 a unit. We substitute this in the second factorization in (1) and cancel p_1 to obtain

$$p_2 p_3 \cdots p_s = u_1 p_2' p_3' \cdots p_t'$$

Set

$$u p_2' = p_2'', \quad p_3' = p_3'', \quad \cdots, \quad p_t' = p_t''.$$

Then

$$p_2 p_3 \cdots p_s = p_2'' p_3'' \cdots p_t''$$

where the p_i'' are irreducible. By the induction assumption we have $s - 1 = t - 1$ and for a suitable ordering of the p_i'', $p_i'' \sim p_i$. Hence $s = t$ and $p_i' \sim p_i'' \sim p_i$ for $i = 2, \cdots, s$.

EXERCISES

1. Show that $I[\sqrt{-5}]$ satisfies A.

2. Let $\mathfrak{A}$ be the set of expressions $a_1 x^{\alpha_1} + a_2 x^{\alpha_2} + \cdots + a_n x^{\alpha_n}$ where the a_i are arbitrary elements in a field $\mathfrak{F}$ and the α_i are non-negative rational numbers. Define addition in the obvious way and multiplication by means of $x^\alpha x^\beta = x^{\alpha+\beta}$. Show that $\mathfrak{A}$ is a commutative integral domain with an identity. Show that the element x of $\mathfrak{A}$ is not a unit but that this element does not have a factorization into irreducible elements.

3. Show that condition B holds in any Gaussian semi-group.

3. Greatest common divisors. Let a be an element of a Gaussian semi-group $\mathfrak{S}$. By combining the associated irreducible factors in a factorization of a, we obtain a factorization

$$(2) \qquad a = u p_1^{e_1} \cdots p_r^{e_r}$$

in which no two of the irreducible elements $p_1, \cdots, p_r$ are associates, the e_i are positive integers, and u is a unit. It is clear now that the factors of a have the form $u' p_1^{e_1'} p_2^{e_2'} \cdots p_r^{e_r'}$ where u' is a unit and the e_i' are integers such that $0 \leq e_i' \leq e_i$.

It is also easy to see that, if a and b are any two non-units, then we can express them in terms of the same non-associate primes, that is, we can write

$$a = u p_1^{e_1} p_2^{e_2} \cdots p_t^{e_t}, \quad b = v p_1^{f_1} p_2^{f_2} \cdots p_t^{f_t}$$

where u and v are units and the e_i and f_i are ≥ 0. Consider now the element

$$d = p_1^{g_1} p_2^{g_2} \cdots p_t^{g_t}, \quad g_i = \min (e_i, f_i).$$

Clearly $d \,|\, a$ and $d \,|\, b$. Moreover, if $c \,|\, a$ and $c \,|\, b$, then $c = w p_1^{k_1} p_2^{k_2} \cdots p_t^{k_t}$, w a unit and $k_i \leq e_i, f_i$. Hence $k_i \leq g_i$ and $c \,|\, d$. This means that the element d is a greatest common divisor of a and b in the sense of the following

Definition 2. *An element d is a greatest common divisor (g.c.d.) of the elements a, b of $\mathfrak{S}$ if $d \,|\, a$ and $d \,|\, b$ and any element c such that $c \,|\, a$ and $c \,|\, b$ is a divisor of d.*

If d is a g.c.d. of a and b, then so is ud, u a unit. On the other hand if d' is any g.c.d. of a and b, then $d \mid d'$ and $d' \mid d$ so that $d \sim d'$. Thus the g.c.d. is determined to within a unit multiplier. We shall find it convenient to denote any determination of the g.c.d. of a and b by (a,b).

We shall now show that the existence of a greatest common divisor for all pairs of elements in an arbitrary semi-group $\mathfrak{S}$ implies that $\mathfrak{S}$ satisfies condition B. Thus we suppose that $\mathfrak{S}$ is any commutative semi-group with identity and cancellation law such that

C. Every pair of elements a, b in $\mathfrak{S}$ has a g.c.d. in $\mathfrak{S}$.

We wish to show that every irreducible element in $\mathfrak{S}$ is prime. For this purpose we require a number of simple lemmas.

Lemma 1. *If C holds in $\mathfrak{S}$, then any finite number of elements of $\mathfrak{S}$ have a g.c.d.*

Let $a,b,c \in \mathfrak{S}$ and set $r = (a,(b,c))$. Then $r \mid a$ and $r \mid (b,c)$ so that $r \mid b$ and $r \mid c$. Also if $s \mid a,b,c$ then $s \mid a$ and $s \mid (b,c)$ so that $s \mid (a,(b,c))$. This shows that $r = (a,(b,c))$ is a g.c.d. of a,b and c. A similar argument holds for more than three factors. Also it is clear that $((a,b),c)$ is a g.c.d. of a,b and c. This proves

Lemma 2. $(a,(b,c)) \sim ((a,b),c)$.

We prove next

Lemma 3. $c(a,b) \sim (ca, cb)$.

Proof. Write $d = (a,b)$ and $e = (ca,cb)$. Then $cd \mid ca$ and $cd \mid cb$. Hence $cd \mid e$. On the other hand, $ca = ex$ and $cb = ey$ and if $e = cdu$, then

$$ca = cdux, \quad cb = cduy.$$

Hence $a = dux$ and $b = duy$. Thus $du \mid a$ and $du \mid b$. Hence $du \mid d$ and u is a unit. This proves the assertion that $c(a,b) \sim (ca,cb)$.

Lemma 4. *If $(a,b) \sim 1$ and $(a,c) \sim 1$ then $(a,bc) \sim 1$.*

Proof. If $(a,b) \sim 1$, then $(ac,bc) \sim c$. Hence $1 \sim (a,c) \sim (a,(ac,bc)) \sim ((a,ac),bc) \sim (a,bc)$.

Now suppose that p is irreducible and that a and b are elements of $\mathfrak{S}$ such that $p \mid ab$. Since p is irreducible and (p,a) is a divisor of p, either $(p,a) \sim p$ or $(p,a) \sim 1$. Similarly $(p,b) \sim p$ or $(p,b) \sim 1$. Now $(p,a) \sim 1$ and $(p,b) \sim 1$ would contradict $(p,ab) \sim p$ by Lemma 4. Hence either $(p,a) \sim p$ or $(p,b) \sim p$. Thus either $p \mid a$ or $p \mid b$. This proves B. The result of the preceding section now yields the following

Theorem 1. *If $\mathfrak{S}$ is a commutative semi-group with identity and cancellation law and $\mathfrak{S}$ satisfies A and C, then $\mathfrak{S}$ is Gaussian.*

We have seen in the Introduction that the semi-group of positive integers and the domain of integers have the greatest common divisor property C. Also it is clear by consideration of absolute values that A holds in these systems. Hence we see that they are Gaussian.

EXERCISES

1. An element m is called a *least common multiple* (l.c.m.) of the elements a and b if $a \mid m$ and $b \mid m$ and if n is any element such that $a \mid n$ and $b \mid n$, then $m \mid n$. Prove that any two elements of a Gaussian semi-group have a l.c.m.

2. Prove that if $\mathfrak{S}$ is Gaussian and $[a,b]$ denotes a l.c.m. of a and b, then $(a,b)[a,b] \sim ab$. Prove also that $[a,(b,c)] \sim ([a,b],[a,c])$.

3. Prove that, if p is a prime positive integer, then the binomial coefficient $\binom{p}{i} = \dfrac{p!}{i!(p-i)!}$, $1 \leq i \leq p - 1$, is divisible by p. Hence prove that in any commutative ring of characteristic p

$$(3) \qquad\qquad (a + b)^p = a^p + b^p$$

holds for every a and b.

4. Define the *Möbius function* $\mu(n)$ of positive integers by the following rules: (a) $\mu(1) = 1$, (b) $\mu(n) = 0$ if n has a square factor, (c) $\mu(n) = (-1)^s$, s the length of n if n is square-free. Prove that $\mu(n)$ is *multiplicative* in the sense that $\mu(n_1 n_2) = \mu(n_1)\mu(n_2)$ if $(n_1,n_2) = 1$. Also prove that

$$\sum_{d \mid n} \mu(d) = \begin{cases} 1 & \text{if } n = 1 \\ 0 & \text{if } n > 1 \end{cases}.$$

5. Prove the Möbius inversion formula: If $f(n)$ is a function of positive integers with values in a ring and

$$g(n) = \sum_{d \mid n} f(d),$$

then

$$f(n) = \sum_{d \mid n} \mu\left(\frac{n}{d}\right) g(d).$$

6. Prove that, if $\phi(n)$ is the Euler ϕ-function, then

$$\phi(n) = \sum_{d \mid n} \mu\left(\frac{n}{d}\right) d.$$

(Cf. ex. 3, p. 34.)

4. Principal ideal domains. Let $\mathfrak{A}$ be a commutative integral domain with an identity. We have defined the principal ideal (b) to be the smallest ideal in $\mathfrak{A}$ containing the element b. Since $\mathfrak{A}$ has an identity, (b) coincides with the totality of multiples bx of the element b. Now $b \mid a$ means that $a = bc \, \varepsilon \, (b)$ and this is equivalent to the requirement that $(a) \subseteq (b)$. Also we note that, if $(a) = (b)$, then $b \mid a$ and $a \mid b$ so that $a \sim b$. The converse is clear too. Hence we see that b is a proper factor of a if and only if $(a) \subset (b)$. The divisor chain condition A for an integral domain $\mathfrak{A}$ can now be stated as the following chain condition on ideals:

A'. $\mathfrak{A}$ contains no infinite properly ascending chain of ideals $(a_1) \subset (a_2) \subset (a_3) \subset \cdots$.

We shall consider now integral domains $\mathfrak{A}$ (commutative and with 1) that have the property that the only ideals in $\mathfrak{A}$ are the principal ideals. A domain of this type is called a *principal ideal domain*. The result that we wish to establish in this section is that every principal ideal domain is Gaussian.

We first prove A'. Let $(a_1) \subseteq (a_2) \subseteq (a_3) \subseteq \cdots$ be an infinite ascending chain of ideals in $\mathfrak{A}$. Let $\mathfrak{B} = \cup (a_i)$ be the logical sum of the sets (a_i). Then we assert that $\mathfrak{B}$ is an ideal in $\mathfrak{A}$. Thus let $b_1, b_2 \, \varepsilon \, \mathfrak{B}$, say $b_1 \, \varepsilon \, (a_k)$, $b_2 \, \varepsilon \, (a_l)$. We can suppose that $k \leq l$. Then $b_1, b_2 \, \varepsilon \, (a_l)$. Hence $b_1 - b_2$ and b_1x for any x are in (a_l). Hence $b_1 - b_2, b_1x \, \varepsilon \, \mathfrak{B}$. This implies that $\mathfrak{B}$ is an ideal. Now by assumption $\mathfrak{B} = (d)$ where $d \, \varepsilon \, \mathfrak{B}$. Since $d \, \varepsilon \, \mathfrak{B}$, $d \, \varepsilon \, (a_n)$ for some integer n. Hence $\mathfrak{B} = (d) = (a_n)$. Consequently, if $m \geq n$, then $(a_m) \supseteq (a_n) = \mathfrak{B} \supseteq (a_m)$ and $(a_m) = (a_n)$. This proves that $\mathfrak{A}$ contains no properly ascending infinite sequences of ideals.

Next let a and b be any two elements of $\mathfrak{A}$ and let (a,b) now denote the ideal $(a) + (b)$ generated by a and b. This ideal is the totality of elements $ax + by$ where x and y are in $\mathfrak{A}$. Now $(a,b) = (d)$. Since $(d) \supseteq (a)$ and $(d) \supseteq (b)$, $d \mid a$ and $d \mid b$. On the other hand, if $e \mid a$ and $e \mid b$, then $(e) \supseteq (a)$ and $(e) \supseteq (b)$. Hence $(e) \supseteq (d)$ and $e \mid d$. This proves that d is a g.c.d. of a and b. Hence C holds and consequently we have the following

Theorem 2. *Every principal ideal domain is Gaussian.*

We have seen that, if $\mathfrak{F}$ is a field, then $\mathfrak{F}[x]$, x transcendental, is a principal ideal domain (Chapter III, § 6). Hence $\mathfrak{F}[x]$ is Gaussian.

EXERCISES

1. Prove that an element p of a commutative integral domain $\mathfrak{A}$ is a prime if and only if $\mathfrak{A}/(p)$ is an integral domain.

2. Prove that, if p is a prime in a principal ideal domain, then $\mathfrak{A}/(p)$ is a field.

3. Let $\mathfrak{A}$ be a principal ideal domain and let $\mathfrak{B}$ be any commutative integral domain containing $\mathfrak{A}$. Show that, if the elements a, b in $\mathfrak{A}$ have the g.c.d. $d \in \mathfrak{A}$, then d is a g.c.d. of a and b in $\mathfrak{B}$.

4. Let $\mathfrak{F}$ be a finite field containing q elements and let $N(r,q)$ denote the number of irreducible polynomials of degree r in $\mathfrak{F}[x]$. Determine $N(2,q)$ and $N(3,q)$.

5. Prove that, if $\mathfrak{A}$ is a commutative integral domain with an identity that is not a field, then $\mathfrak{A}[x]$ is not a principal ideal domain.

5. Euclidean domains. In the ring of integers I the function $\delta(a) = \mid a \mid$ satisfies the conditions:

1. $\delta(a)$ is a non-negative integer, $\delta(a) = 0$ if and only if $a = 0$.
2. $\delta(ab) = \delta(a)\delta(b)$.
3. If $b \neq 0$ and a is arbitrary, then there exist elements q and r such that $a = bq + r$ where $\delta(r) < \delta(b)$.

A similar function can be defined in any polynomial domain $\mathfrak{F}[x]$, $\mathfrak{F}$ a field and x transcendental. Here we take $\delta(a(x)) = 2^{\deg a(x)}$. Then 1 and 2 are immediate and 3 is equivalent to the existence of the division process considered before. The rings I and $\mathfrak{F}[x]$ are examples of Euclidean domains defined in the following

Definition 3. *A commutative integral domain $\mathfrak{A}$ with an identity is a Euclidean domain if there exists a function $\delta(a)$ defined in $\mathfrak{A}$ and satisfying 1, 2, and 3 above.*

We shall give now another example of a Euclidean domain, namely, $I[\sqrt{-1}\,]$, the totality of complex numbers of the form $m + n\sqrt{-1}$ where m and n are integers. Numbers of this type are called *Gaussian integers*. If $a = m + n\sqrt{-1}$, we set $\delta(a) = |a|^2 = m^2 + n^2$. Then 1 and 2 are clear. Now let a and $b \neq 0$ be in $I[\sqrt{-1}\,]$. The complex number $ab^{-1} = \mu + \nu\sqrt{-1}$ where μ and ν are rational numbers. Now we can find integers u and v such that $|u - \mu| \leq \frac{1}{2}$, $|v - \nu| \leq \frac{1}{2}$. Set $\epsilon = \mu - u$, $\eta = \nu - v$, so that $|\epsilon| \leq \frac{1}{2}$ and $|\eta| \leq \frac{1}{2}$. Then

$$a = b[(u + \epsilon) + (v + \eta)\sqrt{-1}\,]$$

$$= bq + r$$

where $q = u + v\sqrt{-1}$ is in $I[\sqrt{-1}\,]$ and $r = b(\epsilon + \eta\sqrt{-1}\,)$. Since $r = a - bq$, r is in $I[\sqrt{-1}\,]$. Moreover,

$$\delta(r) = |r|^2 = |b|^2(\epsilon^2 + \eta^2) \leq |b|^2(\tfrac{1}{4} + \tfrac{1}{4}) = \tfrac{1}{2}\delta(b).$$

Thus $\delta(r) < \delta(b)$.

The main result about Euclidean domains is the following

Theorem 3. *Every Euclidean domain is a principal ideal domain.*

Proof. Let $\mathfrak{B}$ be any ideal in the Euclidean domain $\mathfrak{A}$. If $\mathfrak{B} = 0$, then $\mathfrak{B} = (0)$. Now let $\mathfrak{B} \neq 0$. Then $\mathfrak{B}$ contains elements for which $\delta > 0$ and since the δ's are non-negative integers there exists a $b \,\varepsilon\, \mathfrak{B}$ such that $0 < \delta(b) \leq \delta(c)$ for every $c \neq 0$ in $\mathfrak{B}$. If c is any element of $\mathfrak{B}$, we can write $c = bq + r$ where $\delta(r) < \delta(b)$. But $r = c - bq \,\varepsilon\, \mathfrak{B}$ since $\mathfrak{B}$ is an ideal. Since $\delta(b)$ is the least positive δ for the non zero elements of $\mathfrak{B}$ and $\delta(r) < \delta(b)$, we conclude that $r = 0$. Thus $c = bq \,\varepsilon\, (b)$. Hence $\mathfrak{B} = (b)$ and this completes the proof.

Since every principal ideal domain is Gaussian, we have the

Corollary. *Every Euclidean domain is Gaussian.* *

* Additional results on Euclidean domains are given in § 10 of Chapter VI.

EXERCISES

1. Prove that $I[\sqrt{2}]$, the set of real numbers of the form $m + n\sqrt{2}$, m and n integers, is Euclidean.

2. Let $\mathfrak{A}$ be the totality of complex numbers $m + n\sqrt{-3}$ where m and n are either both integers or both halves of odd integers. Show that $\mathfrak{A}$ is a ring relative to the usual addition and multiplication. Prove that $\mathfrak{A}$ is Euclidean.

3. Prove that an element a of a Euclidean domain is a unit if and only if $\delta(a) = 1$.

4. Let $\mathfrak{A}$ be a Euclidean domain whose function satisfies the condition: $\delta(a + b) \leq \max(\delta(a), \delta(b))$. Show that $\mathfrak{A}$ is either a field or a polynomial domain $\mathfrak{F}[x]$ over a field $\mathfrak{F}$.

6. Polynomial extensions of Gaussian domains. In this section we prove the important theorem that, if $\mathfrak{A}$ is Gaussian and x is transcendental, then $\mathfrak{A}[x]$ is Gaussian.

Let $f(x) = a_0 + a_1 x + \cdots + a_n x^n \neq 0$ be in $\mathfrak{A}[x]$ and let d be the g.c.d. of the non-zero coefficients a_i. We write $a_i = da_i'$ and hence $f(x) = df_1(x)$ where

$$f_1(x) = a_0' + a_1'x + \cdots + a_n'x^n.$$

Evidently the g.c.d. of the non-zero a_i' is 1 (or a unit). A polynomial having this property is called *primitive*. Suppose now that $f(x) = ef_2(x)$ is any factorization of $f(x)$ as a product of a constant e (= element of $\mathfrak{A}$) and a primitive polynomial. Then e is a common factor of the coefficients of $f(x)$ so that $e \mid d$, say $d = ek$. Then $f_2(x) = kf_1(x)$ and, since $f_2(x)$ is primitive, k is a unit. Thus any non-zero polynomial can be written in essentially only one way as a product of a constant and a primitive polynomial.

In studying $\mathfrak{A}[x]$ we find it convenient to introduce the polynomial ring $\mathfrak{F}[x]$ where $\mathfrak{F}$ is the field of fractions of $\mathfrak{A}$. We now prove the following

Lemma 1. *If $f_1(x)$ and $f_2(x)$ are primitive in $\mathfrak{A}[x]$ and are associates in $\mathfrak{F}[x]$, then $f_1(x)$ and $f_2(x)$ are associates in $\mathfrak{A}[x]$.*

Proof. Since the units of $\mathfrak{F}[x]$ are the non-zero elements of $\mathfrak{F}$, we have $f_1(x) = \alpha f_2(x)$, $\alpha \neq 0$ in $\mathfrak{F}$. Write $\alpha = d_2 d_1^{-1}$, d_i in $\mathfrak{A}$. Then $d_1 f_1(x) = d_2 f_2(x)$. This gives two representations of a polynomial in $\mathfrak{A}[x]$ as a product of a constant and a primitive polynomial. It follows that d_1 and d_2 differ by a unit in $\mathfrak{A}$ and that $f_1(x)$ and $f_2(x)$ differ by a unit in $\mathfrak{A}[x]$.

The key result needed to prove that $\mathfrak{A}[x]$ is Gaussian is the following

Lemma 2 (Gauss). *The product of primitive polynomials is primitive.*

Proof. Let $f(x) = a_0 + a_1x + \cdots + a_nx^n$ and $g(x) = b_0 + b_1x + \cdots + b_mx^m$ be primitive and suppose that $f(x)g(x) = c_0 + c_1x + \cdots + c_{n+m}x^{m+n}$ is not primitive. Then there exists an irreducible element $p \in \mathfrak{A}$ such that $p \mid c_i$ for all i. Since $f(x)$ is primitive, p is not a factor of all the a_i and we suppose that $a_{n'}$ is the last a_i not divisible by p. Similarly let $b_{m'}$ be the last b_i not divisible by p. We now consider the coefficient

$$c_{m'+n'} = a_0b_{m'+n'} + a_1b_{m'+n'-1} + \cdots + a_{n'-1}b_{m'+1} + a_{n'}b_{m'}$$
$$+ a_{n'+1}b_{m'-1} + \cdots + a_{n'+m'}b_0.$$

Since all the b_i before the term $a_{n'}b_{m'}$ are divisible by p and since all the a_j after this term are divisible by p and since $c_{m'+n'}$ is divisible by p, $p \mid a_{n'}b_{m'}$. But p is not a divisor of $a_{n'}$ or of $b_{m'}$ and this contradicts the fact that p is irreducible and hence prime (cf. ex. 3, p. 118).

A consequence of Gauss' lemma is

Lemma 3. *If $f(x)$ is an irreducible polynomial of degree > 0 in $\mathfrak{A}[x]$, $f(x)$ is irreducible in $\mathfrak{F}[x]$.*

Proof. Since $f(x)$ is irreducible, it is primitive. Now let $f(x)$ be any primitive polynomial in $\mathfrak{A}[x]$ and suppose that, in $\mathfrak{F}[x]$, $f(x) = \phi_1(x)\phi_2(x)$ where $\deg \phi_i(x) > 0$. Now if $\phi(x)$ is any polynomial $\neq 0$ in $\mathfrak{F}[x]$, let the coefficients of $\phi(x)$ be $\alpha_j = a_jb_j^{-1}$, a_j, b_j in $\mathfrak{A}$. Then we can set

$$\alpha_j = (a_jb_0 \cdots b_{j-1}b_{j+1} \cdots b_n)(b_0b_1 \cdots b_n)^{-1}$$

and this gives us a way of writing the α_j with the same denominator $b = b_0b_1 \cdots b_n$. Thus $\phi(x) = b^{-1}g(x)$ where $g(x) \in \mathfrak{A}[x]$. Also we can write $g(x) = ch(x)$ where $c \in \mathfrak{A}$ and $h(x)$ is primitive. Then $\phi(x) = b^{-1}ch(x)$. We apply these considerations to the $\phi_i(x)$ and obtain $\phi_i(x) = b_i^{-1}c_ih_i(x)$. Then

$$f(x) = b_1^{-1}b_2^{-1}c_1c_2h_1(x)h_2(x)$$

and

$$b_1b_2f(x) = c_1c_2h_1(x)h_2(x).$$

Since the $h_i(x)$ are primitive, $h_1(x)h_2(x)$ is primitive. Hence $f(x) \sim h_1(x)h_2(x)$ and we can suppose that $f(x) = h_1(x)h_2(x)$. Since $\deg h_i(x) = \deg \phi_i(x) > 0$, this is a proper factorization of $f(x)$ in $\mathfrak{A}[x]$. It follows therefore that, if $f(x)$ is irreducible in $\mathfrak{A}[x]$, then it remains irreducible in $\mathfrak{F}[x]$.

We can now prove the main result.

Theorem 4. *If $\mathfrak{A}$ is Gaussian, then so is $\mathfrak{A}[x]$, x transcendental over $\mathfrak{A}$.*

Proof. Let $f(x)$ be $\neq 0$ and $\neq$ a unit. Then $f(x) = df_1(x)$ where $f_1(x)$ is primitive and d is a constant. If $f_1(x)$ is not a unit and is reducible, $f_1(x) = f_{11}(x)f_{12}(x)$. Evidently the $f_{1i}(x)$ have positive degree. Hence $\deg f_{1i}(x) < \deg f_1(x)$. Continuing in this way we arrive at a factorization of $f_1(x)$ as

$$f_1(x) = q_1(x)q_2(x) \cdots q_h(x)$$

where the $q_i(x)$ are irreducible and of positive degree. Also we can factor $d = p_1p_2 \cdots p_s$ where the p_i are irreducible in $\mathfrak{A}$ and hence in $\mathfrak{A}[x]$. This gives a factorization of $f(x)$ into irreducible factors in $\mathfrak{A}[x]$. Now suppose that

$$(4) \qquad \begin{aligned} f(x) &= p_1p_2 \cdots p_s q_1(x)q_2(x) \cdots q_h(x) \\ &= p_1'p_2' \cdots p_i'q_1'(x)q_2'(x) \cdots q_k'(x) \end{aligned}$$

are two factorizations of $f(x)$ into irreducible factors and suppose that the notation has been chosen so that $\deg q_i(x) > 0$, $\deg q_i'(x) > 0$, p_i, $p_i' \in \mathfrak{A}$. Then the $q_i(x)$ and $q_i'(x)$ are primitive. Hence $q_1(x)q_2(x) \cdots q_h(x)$ and $q_1'(x)q_2'(x) \cdots q_k'(x)$ are primitive. It follows that these two products are associates, and, by changing one of the terms by a unit, we can suppose that $\Pi q_i(x) = \Pi q_i'(x)$. Then also $\Pi p_j = \Pi p_j'$. By Lemma 3 the $q_i(x)$ and $q_i'(x)$ are irreducible in $\mathfrak{F}[x]$. Since $\mathfrak{F}[x]$ is Gaussian, the $q_i'(x)$ can be arranged so that $q_i'(x)$ is an associate of $q_i(x)$ in $\mathfrak{F}[x]$. But then Lemma 1 shows that these polynomials are also associates in $\mathfrak{A}[x]$. Finally, since $\mathfrak{A}$ is Gaussian, the primes p_i and p_i' in the factorizations $\Pi p_i = \Pi p_i'$ can be paired off into associate pairs. Hence the two factorizations in (4) are essentially the same.

An immediate corollary of this theorem is that, if $\mathfrak{A}$ is Gaussian and the x_i are algebraically independent, then $\mathfrak{A}[x_1, x_2, \cdots, x_r]$ is Gaussian. For example, if $\mathfrak{F}$ is any field, then $\mathfrak{F}[x_1, x_2, \cdots, x_r]$ is Gaussian. Also $I[x_1, x_2, \cdots, x_r]$ is Gaussian. The rings $\mathfrak{F}[x_1, x_2, \cdots, x_r]$ with $r > 1$ and $I[x_1, x_2, \cdots, x_r]$ with $r \geq 1$ are not principal ideal rings. Hence the class of Gaussian domains is more extensive than the class of principal ideal domains.

EXERCISES

1. Prove that, if $f(x)$ in $I[x]$ has leading coefficient 1 and has a rational root, then this root is an integer.

2. Prove the following irreducibility criterion due to Eisenstein: If $f(x) = a_0 + a_1 x + \cdots + a_n x^n \, \varepsilon \, I[x]$ is primitive and there exists a prime p in I such that $p \mid a_0, p \mid a_1, \cdots, p \mid a_{n-1}$ but $p \nmid a_n$ (p is not a factor of a_n) and $p^2 \nmid a_0$ then $f(x)$ is irreducible in $I[x]$ and hence in $R_0[x]$, R_0 the field of rational numbers.

3. Show that if p is a prime then the polynomial obtained by replacing x by $x + 1$ in $x^{p-1} + x^{p-2} + \cdots + 1 = (x^p - 1)/(x - 1)$ is irreducible in $R_0[x]$. Hence prove that the cyclotomic polynomial $x^{p-1} + x^{p-2} + \cdots + 1$ is irreducible in $R_0[x]$.

Chapter V

GROUPS WITH OPERATORS

In this chapter we resume our study of the theory of groups. The results that we obtain concern the correspondence between the subgroups of a group and those of a homomorphic image, normal series and composition series, the Schreier theorem, direct products and the Krull-Schmidt theorem. The range of application of these results is enormously extended by introducing the new concept of a group with operators. This concept, which was first considered by Krull and by Emmy Noether, enables one to study a group relative to an arbitrary set of endomorphisms. In this way, one achieves a uniform derivation of a number of classical results that were formerly derived separately. Also applications to the theory of rings are obtained by considering the additive group relative to the sets of multiplications as operator domains.

1. Definition and examples of groups with operators

Definition 1. *A* group with operators *is a system consisting of a group $\mathfrak{G}$, a set M and a function defined in the product set $\mathfrak{G} \times M$ and having values in $\mathfrak{G}$ such that, if am denotes the element in $\mathfrak{G}$ determined by the element a of $\mathfrak{G}$ and the element m of M, then*

$$(1) \qquad (ab)m = (am)(bm)$$

holds for any a,b in $\mathfrak{G}$.

If m is fixed and x varies over $\mathfrak{G}$, then $x \rightarrow xm$ is a mapping of $\mathfrak{G}$ into itself. We denote this mapping as $\bar{m}$ and we note that the assumption (1) states that $\bar{m}$ is an endomorphism in $\mathfrak{G}$. Thus every element $m \, \varepsilon \, M$ determines an endomorphism $\bar{m}$ and we have a mapping $m \to \bar{m}$ of M into the set $\mathfrak{E}$ of endomorphisms of $\mathfrak{G}$. It is not required that this mapping be 1–1, that is, we may have $\bar{m} = \bar{n}$ though m and n are distinct in M. These remarks lead to an alternative definition of the concept of a group with operators, namely, the following

Definition 1′. *A* group with operators *is a system consisting of a group $\mathfrak{G}$, a set M and a mapping $m \to \bar{m}$ of M into the set of endomorphisms of $\mathfrak{G}$.*

We have seen that if $\mathfrak{G}, M$ and the mapping $(a,m) \to am$ is a group with operators in the sense of definition 1, then $x \to xm$ is an endomorphism $\bar{m}$ in $\mathfrak{G}$. Also we have the correspondence $m \to \bar{m}$. Hence we have a system satisfying definition 1′. On the other hand, if we have a system of the latter type, then we can define the mapping $(a,m) \to am = a\bar{m}$, and we see that (1) holds. Hence we obtain a group with operators in the original sense. Finally, it is clear that, if we begin with a system satisfying 1 (1′) and we apply successively the two procedures for changing to a system of the other type, then we return to the original system. Hence the two definitions are equivalent.

The second formulation is well suited for constructing examples of groups with operators. For this purpose we can select any set M of endomorphisms of a group $\mathfrak{G}$ and we can let our mapping $m \to \bar{m}$ be the identity. Important sets of endomorphisms that can be used in this way are (1) $\mathfrak{J}$, the set of inner automorphisms, (2) $\mathfrak{A}$, the complete set of automorphisms, (3) $\mathfrak{E}$, the set of endomorphisms.

An example that is conveniently defined by means of the first formulation is the following: $\mathfrak{G}$, the group of vectors in three-dimensional space; M, the set of real numbers; the product function vt for v in $\mathfrak{G}$ and t in M, as the usual product of a vector by a number. Thus, if $v = (x,y,z)$, then

$$vt = (tx,ty,tz).$$

The well-known rule

$$(v + v')t = vt + v't$$

is our requirement (1) in additive dress.

The theory of groups with operators also has important applications to the theory of rings. These applications result in considering certain groups with operators defined in the additive group of a ring. There are three such groups with operators. In all three, the group $\mathfrak{G}$ is the additive group $\mathfrak{A},+$, M is a set of endomorphisms of $\mathfrak{A},+$ and the mapping of M is the identity. In the first case we take $M = \mathfrak{A}_r$, the set of right multiplications. Next we set $M = \mathfrak{A}_l$, the set of left multiplications, and finally we set $M = \mathfrak{A}_r \cup \mathfrak{A}_l$. Accordingly we say that $\mathfrak{A}$ *acts on the right, on the left,* or *on both sides* in its additive group.

We shall usually use the phrase "$\mathfrak{G}$ is a group with operator set M" or "$\mathfrak{G}$ is an M-group" in referring to a group with operators.

We can derive some elementary properties of the product am by using the fact that $\bar{m}$ is an endomorphism. Thus it is clear that $1m = 1$, that $a^{-1}m = (am)^{-1}$ and, more generally, $a^k m = (am)^k$ for any integer k.

2. M-subgroups, M-factor groups and M-homomorphisms. The concept of a group with operators is formulated to focus attention on the collection of subgroups that are sent into themselves by a particular set of endomorphisms; for in studying an M-group it is natural to restrict one's attention to these subgroups of $\mathfrak{G}$. A subgroup $\mathfrak{H}$ is said to be an *M-subgroup* if $hm \, \varepsilon \, \mathfrak{H}$ for every $h \, \varepsilon \, \mathfrak{H}$ and every $m \, \varepsilon \, M$.

It is interesting to see what are the M-subgroups in the examples given in the preceding section. In (1) $M = \mathfrak{J}$ and $\mathfrak{H}$ is an M-subgroup if and only if $g^{-1}\mathfrak{H}g \subseteq \mathfrak{H}$ for every $g \, \varepsilon \, \mathfrak{G}$. Thus the M-subgroups are just the invariant subgroups of $\mathfrak{G}$. In (2) $M = \mathfrak{A}$ and an M-subgroup $\mathfrak{H}$ is, in particular, invariant. Moreover, $\mathfrak{H}$ is mapped into itself by every automorphism of $\mathfrak{G}$. Subgroups having this property are called *characteristic subgroups.* In (3) $M = \mathfrak{E}$, and here $\mathfrak{H}$ is an M-subgroup if and only if $\mathfrak{H}$ is mapped into itself by every endomorphism of $\mathfrak{G}$. Subgroups with this property are said to be *fully invariant.* In the example of the vector group, a subgroup $\mathfrak{H}$ is an M-subgroup if it is closed under scalar multiplication. Such subgroups are called *subspaces.*

We consider also the groups with operators determined by a ring. If $\mathfrak{A}$ acts on the right ($M = \mathfrak{A}_r$), then a subset $\mathfrak{B}$ is an

M-subgroup if and only if it is a subgroup of the additive group $\mathfrak{A}, +$ and it is closed under right multiplication by arbitrary elements of $\mathfrak{A}$. Thus the M-subgroups in this case are the right ideals of the ring. Similarly, if $\mathfrak{A}$ acts on the left, then the M-subgroups are the left ideals. Finally, if $\mathfrak{A}$ acts on both sides, then the M-subgroups are the two-sided ideals.

It is immediate that, if $\{\mathfrak{H}\}$ is a collection of M-subgroups of $\mathfrak{G}$, then the intersection $\cap \mathfrak{H}$ of all these groups is an M-subgroup. Also the group $\mathfrak{S} = [\cup \mathfrak{H}]$ generated by these subgroups is an M-subgroup; for the elements of this group are finite products $h = h_1 h_2 \cdots h_n, h_i \,\varepsilon\, \mathfrak{H}_i \,\varepsilon\, \{\mathfrak{H}\}$. Hence $hm = (h_1 m)(h_2 m) \cdots (h_n m)$ $\varepsilon \, \mathfrak{S}$ since $h_i m \,\varepsilon\, \mathfrak{H}_i$.

If $\mathfrak{H}$ is an M-subgroup of an M-group $\mathfrak{G}$, we can regard $\mathfrak{H}$ as an M-group too. Here we take the product hm, $h \,\varepsilon\, \mathfrak{H}$, $m \,\varepsilon\, M$ to be the product as defined in the M-group $\mathfrak{G}$. Then it is clear that (1) holds. We shall now show that, if $\mathfrak{H}$ is invariant, then there is also a natural way of regarding the factor group $\bar{\mathfrak{G}} = \mathfrak{G}/\mathfrak{H}$ as an M-group. This is done by defining

$$(2) \qquad\qquad (g\mathfrak{H})m = (gm)\mathfrak{H}$$

for every $g \,\varepsilon\, \mathfrak{G}$ and every $m \,\varepsilon\, M$. It is necessary to show that the product thus defined is single-valued and that (1) holds. Now let $g\mathfrak{H} = g'\mathfrak{H}$. Then $g' = gh$, h in $\mathfrak{H}$ and $g'm = (gm)(hm)$ where $hm \,\varepsilon\, \mathfrak{H}$. Hence $(gm)\mathfrak{H} = (g'm)\mathfrak{H}$ and this proves the first assertion. To prove the second we note that

$$((g_1\mathfrak{H})(g_2\mathfrak{H}))m = (g_1 g_2 \mathfrak{H})m = ((g_1 g_2)m)\mathfrak{H} = (g_1 m)(g_2 m)\mathfrak{H}$$

$$= ((g_1 m)\mathfrak{H})((g_2 m)\mathfrak{H}).$$

We shall refer to the group with operators thus defined as the *M-factor group* $\mathfrak{G}/\mathfrak{H}$.

In comparing groups with operators we shall restrict our attention to groups that have the same set of operators M. The basic concept that we consider is that of homomorphism. A mapping η of the M-group $\mathfrak{G}$ into the M-group $\mathfrak{G}'$ is called a *homomorphism* (*M-homomorphism*) if η is a group homomorphism and

$$(3) \qquad\qquad (am)\eta = (a\eta)m$$

holds for all $a \ \varepsilon \ \mathfrak{G}$ and all $m \ \varepsilon \ M$. We have the usual special cases of homomorphism: *isomorphism* if η is 1–1, *endomorphism* if $\mathfrak{G}' = \mathfrak{G}$, *automorphism* if $\mathfrak{G}' = \mathfrak{G}$ and η is 1–1 of $\mathfrak{G}$ onto itself. If there exists an isomorphism of $\mathfrak{G}$ onto $\mathfrak{G}'$, then these M-groups are said to be *isomorphic* ($\cong$).

If η is an M-endomorphism of $\mathfrak{G}$, the condition (3) is equivalent to $\bar{m}\eta = \eta\bar{m}$. Thus the M-endomorphisms are just the endomorphisms that commute with the endomorphisms $\bar{m}$.

Now let η be an M-homomorphism of $\mathfrak{G}$ into $\mathfrak{G}'$ and let $a\eta$ be any element of the image set $\mathfrak{G}\eta$. If $m \ \varepsilon \ M$, $(a\eta)m = (am)\eta \ \varepsilon \ \mathfrak{G}\eta$. Since $\mathfrak{G}\eta$ is a subgroup, this shows that $\mathfrak{G}\eta$ is an M-subgroup of $\mathfrak{G}'$. We consider next the kernel $\mathfrak{K}$ of η. We know that $\mathfrak{K}$ is an invariant subgroup of $\mathfrak{G}$. Also if $k \ \varepsilon \ \mathfrak{K}$ and $m \ \varepsilon \ M$, then $(km)\eta = (k\eta)m = 1'm = 1'$. Hence $km \ \varepsilon \ \mathfrak{K}$ and $\mathfrak{K}$ is an M-subgroup of $\mathfrak{G}$. This proves

Theorem 1. *If η is a homomorphism of the M-group $\mathfrak{G}$ into the M-group $\mathfrak{G}'$, then the image $\mathfrak{G}\eta$ is an M-subgroup of $\mathfrak{G}'$ and the kernel of the homomorphism is an invariant M-subgroup of $\mathfrak{G}$.*

EXERCISES

1. Show that any characteristic (fully invariant) subgroup $\mathfrak{K}$ of a characteristic (fully invariant) subgroup $\mathfrak{H}$ of $\mathfrak{G}$ is characteristic (fully invariant) in $\mathfrak{G}$.

2. Prove that any subgroup of a cyclic group is fully invariant.

3. Show that the subgroup $\mathfrak{G}^{(1)}$ generated by all the *commutators* $[s,t] \equiv sts^{-1}t^{-1}$, s,t in $\mathfrak{G}$, is a fully invariant subgroup. $\mathfrak{G}^{(1)}$ is called the (first) *commutator group* of $\mathfrak{G}$. Prove that $\mathfrak{G}/\mathfrak{G}^{(1)}$ is commutative and that if $\mathfrak{H}$ is any invariant subgroup such that $\mathfrak{G}/\mathfrak{H}$ is commutative then $\mathfrak{H} \supseteq \mathfrak{G}^{(1)}$.

4. Let $\mathfrak{A}$ be a ring with an identity, and regard $\mathfrak{A}$ as an M-group with $M = \mathfrak{A}_r$. What are the M-endomorphisms of $\mathfrak{A}$? Answer the same question for $M = \mathfrak{A}_r \cup \mathfrak{A}_l$.

3. The fundamental theorem of homomorphism for M-groups. It is clear that the resultant of M-homomorphisms is an M-homomorphism. Moreover, if $\mathfrak{H}$ is an invariant M-subgroup of the M-group $\mathfrak{G}$, then the natural mapping ν of $\mathfrak{G}$ onto the M-group $\bar{\mathfrak{G}} = \mathfrak{G}/\mathfrak{H}$ is an M-homomorphism; for by definition $(g\mathfrak{H})m = (gm)\mathfrak{H}$ and, since $g\nu = g\mathfrak{H}$, this means that $g\nu m = gm\nu$.

Next let η be an M-homomorphism of $\mathfrak{G}$ into $\mathfrak{G}'$ and let $\mathfrak{H}$ be an invariant M-subgroup of $\mathfrak{G}$ contained in the kernel $\mathfrak{K}$ of η.

Then as in the case of ordinary groups (cf. p. 44) the correspondence $g\mathfrak{H} \to g\eta$ is single-valued and it defines a homomorphism $\bar{\eta}$ of the M-group $\bar{\mathfrak{G}} = \mathfrak{G}/\mathfrak{H}$ into $\mathfrak{G}'$. The only new fact that has to be established is that $\bar{\eta}$ behaves properly relative to the elements in M, that is, that $((g\mathfrak{H})m)\bar{\eta} = ((g\mathfrak{H})\bar{\eta})m$. This follows from

$$((g\mathfrak{H})m)\bar{\eta} = ((gm)\mathfrak{H})\bar{\eta} = (gm)\eta = (g\eta)m = ((g\mathfrak{H})\bar{\eta})m.$$

As usual we have the factorization $\eta = \nu\bar{\eta}$ where ν is the natural mapping of $\mathfrak{G}$ onto $\bar{\mathfrak{G}}$. Also $\bar{\eta}$ is 1–1 if and only if $\mathfrak{K} = \mathfrak{H}$. This leads immediately to

The fundamental theorem of homomorphism for M-groups. *Any factor group of $\mathfrak{G}$ relative to an invariant M-subgroup is a homomorphic image of $\mathfrak{G}$. Conversely if $\mathfrak{G}'$ is an M-group which is a homomorphic image of the M-group $\mathfrak{G}$, then $\mathfrak{G}'$ is isomorphic to a factor group of $\mathfrak{G}$ relative to an invariant M-subgroup.*

4. The correspondence between M-subgroups determined by a homomorphism. Thus far we have considered only extensions to M-groups of results obtained previously for ordinary groups. We shall begin now to derive some new results. It should be noted that these will apply also to ordinary groups, since the theory of these groups is the special case of the theory of M-groups obtained by taking M to be a vacuous set. Then M-subgroups become ordinary subgroups, M-homomorphisms, ordinary homomorphisms, etc.

Let η be an M-homomorphism of $\mathfrak{G}$ onto $\mathfrak{G}'$ and let $\mathfrak{K}$ be the kernel. If $\mathfrak{H}$ is an M-subgroup of $\mathfrak{G}$, η maps $\mathfrak{H}$ homomorphically onto the M-subgroup $\mathfrak{H}\eta$ of $\mathfrak{G}'$. On the other hand, if $\mathfrak{H}'$ is any M-subgroup of $\mathfrak{G}'$, then the inverse image $\mathfrak{H} = \eta^{-1}(\mathfrak{H}')$ is an M-subgroup of $\mathfrak{G}$; for, if $h_1, h_2 \,\varepsilon\, \mathfrak{H}$, then $(h_1h_2^{-1})\eta = (h_1\eta)(h_2\eta)^{-1} \,\varepsilon\, \mathfrak{H}'$ so that $h_1h_2^{-1} \,\varepsilon\, \mathfrak{H}$. Also if $h \,\varepsilon\, \mathfrak{H}$ and $m \,\varepsilon\, M$, then $(hm)\eta = (h\eta)m \,\varepsilon\, \mathfrak{H}'$. Hence $hm \,\varepsilon\, \mathfrak{H}$.

Evidently $\mathfrak{H} = \eta^{-1}(\mathfrak{H}')$ contains $\mathfrak{K} = \eta^{-1}(1')$ and $\mathfrak{H}\eta = \mathfrak{H}'$. Thus we see that we can obtain every M-subgroup of $\mathfrak{G}'$ by applying η to an M-subgroup of $\mathfrak{G}$ that contains $\mathfrak{K}$. Now let $\mathfrak{H}$ be any M-subgroup of $\mathfrak{G}$ that contains $\mathfrak{K}$ and let $\mathfrak{H}_1 = \eta^{-1}(\mathfrak{H}\eta)$. Clearly $\mathfrak{H}_1 \supseteq \mathfrak{H}$. On the other hand, if $h_1 \,\varepsilon\, \mathfrak{H}_1$, then $h_1\eta = h\eta$ for some

h in $\mathfrak{H}$. Hence $h_1 = hk$, k in $\mathfrak{K}$. Since $\mathfrak{H} \supseteq \mathfrak{K}$, this implies that $h_1 \,\varepsilon\, \mathfrak{H}$. Hence $\eta^{-1}(\mathfrak{H}\eta) = \mathfrak{H}$.

We can now easily prove the following

Theorem 2. *Let η be an M-homomorphism of $\mathfrak{G}$ onto $\mathfrak{G}'$ with kernel $\mathfrak{K}$ and let $\{\mathfrak{H}\}$ be the collection of M-subgroups of $\mathfrak{G}$ that contain $\mathfrak{K}$. Then the mapping $\mathfrak{H} \to \mathfrak{H}\eta$ is 1-1 of $\{\mathfrak{H}\}$ onto the collection of M-subgroups of $\mathfrak{G}'$. The subgroup $\mathfrak{H}$ is invariant in $\mathfrak{G}$ if and only if its image $\mathfrak{H}' = \mathfrak{H}\eta$ is invariant in $\mathfrak{G}'$.*

Proof. We have seen that $\mathfrak{H} \to \mathfrak{H}\eta$ is a mapping of $\{\mathfrak{H}\}$ onto the set of M-subgroups of $\mathfrak{G}'$. Also if $\mathfrak{H}_1$ and $\mathfrak{H}_2 \,\varepsilon\, \{\mathfrak{H}\}$ and $\mathfrak{H}_1\eta = \mathfrak{H}_2\eta$, then $\mathfrak{H}_1 = \eta^{-1}(\mathfrak{H}_1\eta) = \eta^{-1}(\mathfrak{H}_2\eta) = \mathfrak{H}_2$. Hence, our mapping is 1-1. It is easy to verify that $\mathfrak{H}$ is invariant in $\mathfrak{G}$ if and only if $\mathfrak{H}' = \mathfrak{H}\eta$ is invariant in $\mathfrak{G}'$.

An important special case of this theorem is obtained by considering the natural homomorphism v of $\mathfrak{G}$ onto an M-factor group $\mathfrak{G}/\mathfrak{K}$, $\mathfrak{K}$ an invariant M-subgroup. In this case, we see that any M-subgroup of $\bar{\mathfrak{G}} = \mathfrak{G}/\mathfrak{K}$ is obtained by applying v to an M-subgroup $\mathfrak{H}$ of $\mathfrak{G}$ that contains $\mathfrak{K}$. The image $\mathfrak{H}v$ is the set of cosets $h\mathfrak{K}$, $h \,\varepsilon\, \mathfrak{H}$; hence it is just the factor group $\mathfrak{H}/\mathfrak{K}$. We can therefore state the following

Corollary. *Let $\mathfrak{G}$ be an M-group and $\mathfrak{K}$ an invariant M-subgroup. Then any M-subgroup of the M-factor group $\mathfrak{G}/\mathfrak{K}$ has the form $\mathfrak{H}/\mathfrak{K}$ where $\mathfrak{H}$ is an M-subgroup of $\mathfrak{G}$ containing $\mathfrak{K}$. Distinct $\mathfrak{H}$'s give rise in this way to distinct M-subgroups of $\mathfrak{G}/\mathfrak{K}$, and $\mathfrak{H}$ is invariant in $\mathfrak{G}$ if and only if $\mathfrak{H}/\mathfrak{K}$ is invariant in $\mathfrak{G}/\mathfrak{K}$.*

Analogous results can be proved for rings. These can either be proved directly, or they can be obtained as special cases of the group theorems. We shall employ the second method here. Let η be a homomorphism of the ring $\mathfrak{A}$ onto the ring $\mathfrak{A}'$ and let $\mathfrak{K}$ be the kernel of η. Then we can consider $\mathfrak{A},+$ as a group with the operator set $M = \mathfrak{A}_r \cup \mathfrak{A}_l$. Moreover, we can also consider $\mathfrak{A}',+$ as an M-group; for we can define

$$x'a_r \equiv x'(a\eta)_r = x'(a\eta)$$

$$\text{(4)}$$

$$x'a_l \equiv x'(a\eta)_l = (a\eta)x',$$

and it is clear that the basic requirement (1) is fulfilled. **When** this definition is used, η becomes an M-homomorphism of $\mathfrak{A},+$ onto $\mathfrak{A}',+$, since

$$(x\eta)a_r = (x\eta)(a\eta) = (xa)\eta = (xa_r)\eta$$

$$(x\eta)a_l = (a\eta)(x\eta) = (ax)\eta = (xa_l)\eta.$$

Finally we need to observe that the M-subgroups of $\mathfrak{A}',+$ are just the (two-sided) ideals of the ring $\mathfrak{A}'$; for, if $\mathfrak{B}'$ is an M-subgroup, then $b'(a\eta)$ and $(a\eta)b' \; \varepsilon \; \mathfrak{B}'$ for every b' in $\mathfrak{B}'$. Since the set $\{a\eta\} = \mathfrak{A}'$, $\mathfrak{B}'$ is an ideal. The converse is clear, too. Now Theorem 2 establishes a 1–1 correspondence between the set $\{\mathfrak{B}\}$ of ideals of $\mathfrak{A}$ that contain $\mathfrak{K}$ and the complete set of ideals in $\mathfrak{A}'$. In particular, we have a 1–1 correspondence between the set of ideals $\{\mathfrak{B}\}$, $\mathfrak{B} \supseteq \mathfrak{K}$, and the ideals of the difference ring $\mathfrak{A}/\mathfrak{K}$. Any ideal of $\mathfrak{A}/\mathfrak{K}$ has the form $\mathfrak{B}/\mathfrak{K}$, $\mathfrak{B}$ an ideal of $\mathfrak{A}$ containing $\mathfrak{K}$. Distinct $\mathfrak{B}$'s give rise to distinct ideals $\mathfrak{B}/\mathfrak{K}$.

EXERCISES

1. Determine the ideals of $I/(m)$, $m > 0$.
2. Give a direct derivation of the correspondence between ideals of a ring and those of a homomorphic image.

5. The isomorphism theorems for M-groups. In this section we shall prove three important theorems on the isomorphism of M-groups. The first of these can be regarded as a supplement to the theorem establishing the correspondence between the subgroups of a group and of a homomorphic image. As before, let η be a homomorphism of the M-group $\mathfrak{G}$ onto the M-group $\mathfrak{G}'$ and let $\mathfrak{K}$ be the kernel. Let $\mathfrak{H}$ be an invariant M-subgroup of $\mathfrak{G}$ that contains the kernel $\mathfrak{K}$ and let $\mathfrak{H}' = \mathfrak{H}\eta$. Then, if ν' is the natural homomorphism of $\mathfrak{G}'$ onto $\mathfrak{G}'/\mathfrak{H}'$, $\eta\nu'$ is a homomorphism of $\mathfrak{G}$ onto $\mathfrak{G}'/\mathfrak{H}'$. If $g\eta\nu' = \mathfrak{H}'$, $g\eta \; \varepsilon \; \mathfrak{H}'$ and conversely. Hence the kernel of $\eta\nu'$ is the group $\mathfrak{H}$. By the fundamental theorem the mapping $\overline{\eta\nu'}$ defined by $g\mathfrak{H} \to g\eta\nu' = (g\eta)\mathfrak{H}'$ is an M-isomorphism of $\mathfrak{G}/\mathfrak{H}$ onto $\mathfrak{G}'/\mathfrak{H}'$. This proves the

First isomorphism theorem. *Let η be a homomorphism of the M-group $\mathfrak{G}$ onto the M-group $\mathfrak{G}'$ with kernel $\mathfrak{K}$ and let $\mathfrak{H}$ be an invariant M-subgroup of $\mathfrak{G}$ that contains $\mathfrak{K}$. Then $\mathfrak{H}\eta = \mathfrak{H}'$ is*

invariant in $\mathfrak{G}'$ and the M-factor groups $\mathfrak{G}/\mathfrak{H}$ and $\mathfrak{G}'/\mathfrak{H}'$ are isomorphic under the correspondence $g\mathfrak{H} \to (g\eta)\mathfrak{H}'$.

As a special case of this theorem we take $\mathfrak{G}'$ to be the *M*-factor group $\mathfrak{G}/\mathfrak{K}$, and $\eta = \nu$ the natural homomorphism. If $\mathfrak{H}$ is an invariant *M*-subgroup of $\mathfrak{G}$ containing $\mathfrak{K}$, then $\mathfrak{H}\eta$ is the factor group $\mathfrak{H}/\mathfrak{K}$ of cosets $h\mathfrak{K}$, h in $\mathfrak{H}$. Hence we have the

Corollary. *If $\mathfrak{K}$ and $\mathfrak{H}$ are invariant M-subgroups of $\mathfrak{G}$ and $\mathfrak{H} \supseteq \mathfrak{K}$, then $\mathfrak{G}/\mathfrak{H}$ and $(\mathfrak{G}/\mathfrak{K})/(\mathfrak{H}/\mathfrak{K})$ are isomorphic.*

Assume next that $\mathfrak{G}_1$ and $\mathfrak{G}_2$ are *M*-subgroups of $\mathfrak{G}$ and that $\mathfrak{G}_2$ is invariant. The *M*-subgroup generated by $\mathfrak{G}_1$ and $\mathfrak{G}_2$ is the product set $\mathfrak{G}_1\mathfrak{G}_2 = \mathfrak{G}_2\mathfrak{G}_1$. It is clear that the correspondence $g_1 \to g_1\mathfrak{G}_2$, g_1 in $\mathfrak{G}_1$, is a homomorphism of the *M*-subgroup $\mathfrak{G}_1$ into $\mathfrak{G}_1\mathfrak{G}_2/\mathfrak{G}_2$. Any coset in $\mathfrak{G}_1\mathfrak{G}_2$ has the form $g_1g_2\mathfrak{G}_2 = g_1\mathfrak{G}_2$, $g_i \,\varepsilon\, \mathfrak{G}_i$. Hence our homomorphism is a mapping onto $\mathfrak{G}_1\mathfrak{G}_2/\mathfrak{G}_2$. If $g_1\mathfrak{G}_2 = \mathfrak{G}_2$, then $g_1 \,\varepsilon\, \mathfrak{G}_2$ and so $g_1 \,\varepsilon\, \mathfrak{G}_1 \cap \mathfrak{G}_2$. This shows that the kernel of the homomorphism $g_1 \to g_1\mathfrak{G}_2$ is $\mathfrak{G}_1 \cap \mathfrak{G}_2$. We therefore have the following

Second isomorphism theorem. *If $\mathfrak{G}_1$ and $\mathfrak{G}_2$ are M-subgroups of a group and $\mathfrak{G}_2$ is invariant, then* (1) *$\mathfrak{G}_1 \cap \mathfrak{G}_2$ is invariant in $\mathfrak{G}_1$, and* (2) *the M-factor groups $\mathfrak{G}_1\mathfrak{G}_2/\mathfrak{G}_2$ and $\mathfrak{G}_1/(\mathfrak{G}_1 \cap \mathfrak{G}_2)$ are isomorphic under the correspondence $g_1\mathfrak{G}_2 \to g_1(\mathfrak{G}_1 \cap \mathfrak{G}_2)$.*

We shall establish next a somewhat more complicated isomorphism theorem which will be used in the next section to prove an important refinement theorem due to Schreier.

Third isomorphism theorem (Zassenhaus). *Let $\mathfrak{G}_i'$ and $\mathfrak{G}_i$, $i = 1,2$, be M-subgroups of $\mathfrak{G}$ such that $\mathfrak{G}_i'$ is invariant in $\mathfrak{G}_i$. Then $(\mathfrak{G}_1 \cap \mathfrak{G}_2')\mathfrak{G}_1'$ is invariant in $(\mathfrak{G}_1 \cap \mathfrak{G}_2)\mathfrak{G}_1'$, $(\mathfrak{G}_1' \cap \mathfrak{G}_2)\mathfrak{G}_2'$ is invariant in $(\mathfrak{G}_1 \cap \mathfrak{G}_2)\mathfrak{G}_2'$ and the corresponding factor groups are M-isomorphic.*

Proof. Consider the subgroup $(\mathfrak{G}_1 \cap \mathfrak{G}_2')\mathfrak{G}_1'$ of $(\mathfrak{G}_1 \cap \mathfrak{G}_2)\mathfrak{G}_1'$. First we show directly that it is invariant: Let $x \,\varepsilon\, \mathfrak{G}_1 \cap \mathfrak{G}_2$; $y \,\varepsilon\, \mathfrak{G}_1 \cap \mathfrak{G}_2'$; $z,t \,\varepsilon\, \mathfrak{G}_1'$. Then $x^{-1}yx \,\varepsilon\, \mathfrak{G}_1 \cap \mathfrak{G}_2'$ and $x^{-1}zx \,\varepsilon\, \mathfrak{G}_1'$ whence

$$(5) \qquad x^{-1}(\mathfrak{G}_1 \cap \mathfrak{G}_2')\mathfrak{G}_1'x \subseteq (\mathfrak{G}_1 \cap \mathfrak{G}_2')\mathfrak{G}_1'.$$

Also $t^{-1}yt = t^{-1}(yty^{-1})y$, and since $yty^{-1} \, \varepsilon \, \mathfrak{G}_1'$, we have $t^{-1}yt \, \varepsilon$
$\mathfrak{G}_1'(\mathfrak{G}_1 \cap \mathfrak{G}_2') = (\mathfrak{G}_1 \cap \mathfrak{G}_2')\mathfrak{G}_1'$. Hence

$$(6) \quad t^{-1}(\mathfrak{G}_1 \cap \mathfrak{G}_2')\mathfrak{G}_1't = t^{-1}(\mathfrak{G}_1 \cap \mathfrak{G}_2')tt^{-1}\mathfrak{G}_1't \subseteq (\mathfrak{G}_1 \cap \mathfrak{G}_2')\mathfrak{G}_1'.$$

It is clear from (5) and (6) that $(\mathfrak{G}_1 \cap \mathfrak{G}_2')\mathfrak{G}_1'$ is invariant in
$(\mathfrak{G}_1 \cap \mathfrak{G}_2)\mathfrak{G}_1'$. By the second isomorphism theorem, it follows
that $(\mathfrak{G}_1 \cap \mathfrak{G}_2')\mathfrak{G}_1' \cap (\mathfrak{G}_1 \cap \mathfrak{G}_2)$ is invariant in $\mathfrak{G}_1 \cap \mathfrak{G}_2$ and

$$(7) \quad (\mathfrak{G}_1 \cap \mathfrak{G}_2)/(\mathfrak{G}_1 \cap \mathfrak{G}_2')\mathfrak{G}_1' \cap (\mathfrak{G}_1 \cap \mathfrak{G}_2)$$

$$\cong (\mathfrak{G}_1 \cap \mathfrak{G}_2)(\mathfrak{G}_1 \cap \mathfrak{G}_2')\mathfrak{G}_1'/(\mathfrak{G}_1 \cap \mathfrak{G}_2')\mathfrak{G}_1'$$

$$= (\mathfrak{G}_1 \cap \mathfrak{G}_2)\mathfrak{G}_1'/(\mathfrak{G}_1 \cap \mathfrak{G}_2')\mathfrak{G}_1'.$$

On the other hand,

$$(8) \quad (\mathfrak{G}_1 \cap \mathfrak{G}_2')\mathfrak{G}_1' \cap (\mathfrak{G}_1 \cap \mathfrak{G}_2) = (\mathfrak{G}_1 \cap \mathfrak{G}_2')\mathfrak{G}_1' \cap \mathfrak{G}_2$$

and any element of $(\mathfrak{G}_1 \cap \mathfrak{G}_2')\mathfrak{G}_1'$ has the form yz, $y \, \varepsilon \, \mathfrak{G}_1 \cap \mathfrak{G}_2'$,
$z \, \varepsilon \, \mathfrak{G}_1'$. If $yz \, \varepsilon \, \mathfrak{G}_2$ then $z = y^{-1}(yz) \, \varepsilon \, \mathfrak{G}_2$ so that $z \, \varepsilon \, \mathfrak{G}_2 \cap \mathfrak{G}_1'$.
Hence $yz \, \varepsilon \, (\mathfrak{G}_1 \cap \mathfrak{G}_2')(\mathfrak{G}_1' \cap \mathfrak{G}_2)$ and $(\mathfrak{G}_1 \cap \mathfrak{G}_2')\mathfrak{G}_1' \cap \mathfrak{G}_2 \subseteq$
$(\mathfrak{G}_1 \cap \mathfrak{G}_2')(\mathfrak{G}_1' \cap \mathfrak{G}_2)$. The reverse inequality is clear. Hence
$(\mathfrak{G}_1 \cap \mathfrak{G}_2')\mathfrak{G}_1' \cap \mathfrak{G}_2 = (\mathfrak{G}_1 \cap \mathfrak{G}_2')(\mathfrak{G}_1' \cap \mathfrak{G}_2)$. Consequently (7)
can be re-written as

$$(9) \quad (\mathfrak{G}_1 \cap \mathfrak{G}_2)/(\mathfrak{G}_1 \cap \mathfrak{G}_2')(\mathfrak{G}_1' \cap \mathfrak{G}_2)$$

$$\cong (\mathfrak{G}_1 \cap \mathfrak{G}_2)\mathfrak{G}_1'/(\mathfrak{G}_1 \cap \mathfrak{G}_2')\mathfrak{G}_1'.$$

By symmetry we have also

$$(10) \quad (\mathfrak{G}_1 \cap \mathfrak{G}_2)/(\mathfrak{G}_1 \cap \mathfrak{G}_2')(\mathfrak{G}_1' \cap \mathfrak{G}_2)$$

$$\cong (\mathfrak{G}_1 \cap \mathfrak{G}_2)\mathfrak{G}_2'/(\mathfrak{G}_2 \cap \mathfrak{G}_1')\mathfrak{G}_2'.$$

Our result now follows from (9) and (10).

EXERCISES

1. Show that the third isomorphism theorem implies the second.

2. Let $\mathfrak{G}_1$, $\mathfrak{G}_1'$ be M-subgroups such that $\mathfrak{G}_1'$ is invariant in $\mathfrak{G}_1$ and let $\mathfrak{H}$
be any M-subgroup of $\mathfrak{G}$. Prove that $\mathfrak{H}_1' = \mathfrak{G}_1' \cap \mathfrak{H}$ is invariant in $\mathfrak{H}_1 = \mathfrak{G}_1 \cap \mathfrak{H}$ and that $\mathfrak{H}_1/\mathfrak{H}_1'$ is isomorphic to a subgroup of $\mathfrak{G}_1/\mathfrak{G}_1'$.

3. State the ring analogues of the first and second isomorphism theorems.

6. Schreier's theorem. We shall consider now a type of factorization of a group into factor groups. Let

$$(11) \qquad \mathfrak{G} = \mathfrak{G}_1 \supseteq \mathfrak{G}_2 \supseteq \cdots \supseteq \mathfrak{G}_{s+1} = 1$$

be a sequence of M-subgroups of the M-group $\mathfrak{G}$ such that each $\mathfrak{G}_{i+1}$ is invariant in $\mathfrak{G}_i$. We call such a sequence a *normal series* for $\mathfrak{G}$. The factor groups

$$(12) \qquad \mathfrak{G}_1/\mathfrak{G}_2, \quad \mathfrak{G}_2/\mathfrak{G}_3, \quad \cdots, \quad \mathfrak{G}_s/\mathfrak{G}_{s+1} = \mathfrak{G}_s$$

are the *factors* of the normal series. As an example we let $\mathfrak{G}$ be the finite cyclic group of order n. Then the subgroup $\mathfrak{G}_i$ is determined by its order n_i and $n_{i+1} \mid n_i$. The ratio $q_i = n_i/n_{i+1}$ is the order of $\mathfrak{G}_i/\mathfrak{G}_{i+1}$. Since $n = n_1 = q_1 n_2$, $n_2 = q_2 n_3$, $\cdots$, $n = q_1 q_2 \cdots q_s$. Conversely, if $n = q_1 q_2 \cdots q_s$ is a factorization of n, then the cyclic group $\mathfrak{G}$ has a subgroup $\mathfrak{G}_i$ of order $n_i = q_i q_{i+1} \cdots q_s$. Hence $\mathfrak{G} = \mathfrak{G}_1 \supseteq \mathfrak{G}_2 \supseteq \cdots \supseteq \mathfrak{G}_{s+1} = 1$, and the order of $\mathfrak{G}_i/\mathfrak{G}_{i+1}$ is q_i.

The two normal series

$$
(13) \qquad
\begin{aligned}
\mathfrak{G} &= \mathfrak{G}_1 \supseteq \mathfrak{G}_2 \supseteq \cdots \supseteq \mathfrak{G}_{s+1} = 1 \\
\mathfrak{G} &= \mathfrak{H}_1 \supseteq \mathfrak{H}_2 \supseteq \cdots \supseteq \mathfrak{H}_{t+1} = 1
\end{aligned}
$$

are said to be *equivalent* if it is possible to set up a 1–1 correspondence between the factors of the two series such that the paired factors are isomorphic. We say that one normal series is a *refinement* of a second if its terms include all of the groups that occur in the second series. We can now state the following fundamental theorem.

Schreier's refinement theorem. *Any two normal series for an M-group have equivalent refinements.*

Proof. Let the two series be given by (13). We set

$$
(14) \qquad
\begin{aligned}
\mathfrak{G}_{ik} &= (\mathfrak{G}_i \cap \mathfrak{H}_k)\mathfrak{G}_{i+1}, \quad k = 1, 2, \cdots, t+1 \\
\mathfrak{H}_{ki} &= (\mathfrak{G}_i \cap \mathfrak{H}_k)\mathfrak{H}_{k+1}, \quad i = 1, 2, \cdots, s+1.
\end{aligned}
$$

Then

$$
(15) \qquad
\begin{aligned}
\mathfrak{G} = \mathfrak{G}_{11} &\supseteq \mathfrak{G}_{12} \supseteq \cdots \supseteq \mathfrak{G}_{1,t+1} \\
&= \mathfrak{G}_{21} \supseteq \mathfrak{G}_{22} \supseteq \cdots \supseteq \mathfrak{G}_{2,t+1} \cdots \supseteq \mathfrak{G}_{s,t+1} = 1, \\
\mathfrak{G} = \mathfrak{H}_{11} &\supseteq \mathfrak{H}_{12} \supseteq \cdots \supseteq \mathfrak{H}_{1,s+1} \\
&= \mathfrak{H}_{21} \supseteq \mathfrak{H}_{22} \supseteq \cdots \supseteq \mathfrak{H}_{2,s+1} \cdots \supseteq \mathfrak{H}_{t,s+1} = 1.
\end{aligned}
$$

Now we can apply the third isomorphism theorem to the groups $\mathfrak{G}_i$, $\mathfrak{H}_k$, $\mathfrak{G}_{i+1}$, $\mathfrak{H}_{k+1}$ to conclude that $\mathfrak{G}_{i,k+1} = (\mathfrak{G}_i \cap \mathfrak{H}_{k+1})\mathfrak{G}_{i+1}$ is invariant in $\mathfrak{G}_{ik} = (\mathfrak{G}_i \cap \mathfrak{H}_k)\mathfrak{G}_{i+1}$, that $\mathfrak{H}_{k,i+1} = (\mathfrak{G}_{i+1} \cap \mathfrak{H}_k)\mathfrak{H}_{k+1}$ is invariant in $\mathfrak{H}_{ki} = (\mathfrak{G}_i \cap \mathfrak{H}_k)\mathfrak{H}_{k+1}$ and that $\mathfrak{G}_{ik}/\mathfrak{G}_{i,k+1} \cong \mathfrak{H}_{ki}/\mathfrak{H}_{k,i+1}$. Hence the two series in (15) are normal and equivalent. Since these series are refinements of the series given in (13), this proves the theorem.

EXERCISES

1. Show that, if $\mathfrak{G} = \mathfrak{G}_1 \supseteq \mathfrak{G}_2 \supseteq \cdots \supseteq \mathfrak{G}_{s+1} = 1$ is a normal series for $\mathfrak{G}$ and $\mathfrak{H}$ is any M-subgroup, then $\mathfrak{H} = (\mathfrak{H} \cap \mathfrak{G}_1) \supseteq (\mathfrak{H} \cap \mathfrak{G}_2) \supseteq \cdots \supseteq (\mathfrak{H} \cap \mathfrak{G}_{s+1}) = 1$ is a normal series for $\mathfrak{H}$. Show that the factors of the second series are isomorphic to subgroups of the factors of the first series.

2. An ordinary group is called *solvable* if it has a normal series whose factors are commutative groups. Prove that any subgroup and any factor group of a solvable group is solvable.

3. Define the higher derived groups of $\mathfrak{G}$ inductively by $\mathfrak{G}^{(i)} = (\mathfrak{G}^{(i-1)})^{(1)}$ (cf. ex. 3, p. 132). Prove that $\mathfrak{G}$ is solvable if and only if $\mathfrak{G}^{(s)} = 1$ for some integer s.

4. Prove that any finite group of prime power order is solvable (cf. ex. 3, p. 48).

7. Simple groups and the Jordan-Hölder theorem. The subgroups $\mathfrak{G}$ and 1 are invariant M-subgroups in any M-group $\mathfrak{G}$. If $\mathfrak{G} \neq 1$ and these are the only invariant M-subgroups, then $\mathfrak{G}$ is called *M-simple*. For example, any cyclic group of prime order is simple. Another important class of simple groups is furnished by the following

Theorem 3. *The alternating group A_n is simple if $n \geq 5$.*

Proof. We have seen (ex. 2, p. 37) that A_n is generated by its three-cycles $(i\,j\,k)$. We note next that, if an invariant subgroup $\mathfrak{H}$ of A_n contains one three-cycle, then it contains every three-cycle; hence, it coincides with A_n. For let $(1\,2\,3)\ \varepsilon\ \mathfrak{H}$ and let $(i\,j\,k)$ be any three-cycle. Then we can extend the mapping $1 \to i, 2 \to j, 3 \to k$ to a permutation

$$\gamma = \begin{pmatrix} 1 & 2 & 3 & 4 & 5 & \cdots \\ i & j & k & l & m & \cdots \end{pmatrix}$$

of $1, 2, \cdots, n$. If γ is odd, we can multiply it on the right by (lm) to obtain an even permutation. Hence, we may suppose that

$\gamma \varepsilon A_n$. Since $\gamma^{-1}(1\,2\,3)\gamma = (i\,j\,k)\; \varepsilon \; \mathfrak{H}$ this proves our assertion. We shall now show that, if $\mathfrak{H} \neq 1$, then $\mathfrak{H}$ contains a three-cycle. Let α be a permutation belonging to $\mathfrak{H}$ that is $\neq 1$ and that leaves fixed as many elements as any other permutation $\neq 1$ in $\mathfrak{H}$. If α is not a three-cycle, either α contains a cycle of length ≥ 3 and moves more than three elements or α is a product of at least two disjoint transpositions. Accordingly we may assume that either

$$(16) \qquad \alpha = (1\,2\,3\,\cdots)(\quad) \cdots$$

or

$$(17) \qquad \alpha = (1\,2)(3\,4) \cdots .$$

In the first case α moves at least two other numbers, say 4,5, since α is not one of the odd permutations $(1\,2\,3\,k)$. Now let $\beta = (3\,4\,5)$ and form $\alpha_1 = \beta^{-1}\alpha\beta$. If α is as in (16)

$$\alpha_1 = (1\,2\,4\,\cdots)(\quad) \cdots$$

and if α is as in (17)

$$\alpha_1 = (1\,2)(4\,5) \cdots .$$

Now it is clear that, if a number $i > 5$ is left fixed by α, then it is also left fixed by α_1 and hence it is left fixed by $\alpha_1\alpha^{-1}$. Moreover $1\alpha_1\alpha^{-1} = 1$ if α is as in (16) and $1\alpha_1\alpha^{-1} = 1$ and $2\alpha_1\alpha^{-1} = 2$ if α is as in (17). Thus $\alpha_1\alpha^{-1}$ leaves invariant more elements than α. Since $\alpha_1\alpha^{-1} \neq 1$, this contradicts our choice of α. Hence α is a three-cycle, and the theorem is proved.*

We shall say that the invariant M-subgroup $\mathfrak{H}$ of $\mathfrak{G}$ is *maximal* in $\mathfrak{G}$ if $\mathfrak{G} \supset \mathfrak{H}$ and there exists no invariant M-subgroup $\mathfrak{K}$ such that $\mathfrak{G} \supset \mathfrak{K} \supset \mathfrak{H}$. It is clear from our correspondence between subgroups of a group and those of a factor group that $\mathfrak{H}$ *is maximal in $\mathfrak{G}$ if and only if $\mathfrak{G}/\mathfrak{H}$ is M-simple.*

We now define a *composition series* for a group $\mathfrak{G}$ to be a normal series

$$(18) \qquad \mathfrak{G} = \mathfrak{G}_1 \supset \mathfrak{G}_2 \supset \cdots \supset \mathfrak{G}_{s+1} = 1$$

---

*This proof is essentially the same as the one given in van der Waerden's *Moderne Algebra*.

with the property that each $\mathfrak{G}_{i+1}$ is maximal in $\mathfrak{G}_i$. Thus a composition series is a normal series whose factors are simple groups $\neq 1$. An M-group $\mathfrak{G}$ need not have a composition series. For example, if M is vacuous and $\mathfrak{G}$ is an infinite commutative group, then $\mathfrak{G}$ does not have a composition series. To see this we note first that a simple commutative group has no subgroups other than 1 and the whole group. Therefore, such a group is necessarily a finite cyclic group of prime order. Hence, if (18) is a composition series for an ordinary commutative group, then the factor groups $\mathfrak{G}_i/\mathfrak{G}_{i+1}$ are cyclic of prime order. Now if a group $\mathfrak{G}$ contains a subgroup $\mathfrak{H}$ of finite order m and finite index r, then $\mathfrak{G}$ is of finite order mr. It follows easily from this that a group that has a composition series whose factors are finite groups is itself finite. In particular, we see that, if $\mathfrak{G}$ is an ordinary commutative group with a composition series, then $\mathfrak{G}$ is finite.

If an M-group does have a composition series, then the composition factors ($=$ factors of the composition series) are uniquely determined by the group. This is the content of the

Jordan-Hölder theorem. *Any two composition series for an M-group are equivalent.*

Proof. By Schreier's theorem the composition series have equivalent refinements. On the other hand, it is clear from the definition of a composition series that a refinement of such a series has the same factors $\neq 1$ as the given series. Now in the 1–1 correspondence between the factors of the refinements the factors $= 1$ are paired. Hence, the factors $\neq 1$ are also paired. Since these are the composition factors of the given composition series, we see that the two composition series are equivalent.

EXERCISES

1. Apply the Jordan-Hölder theorem for finite cyclic groups to prove the uniqueness of factorization of a positive integer into positive primes.

2. Show that, if $\mathfrak{G}$ has a composition series, then any normal series for $\mathfrak{G}$ in which the terms are properly decreasing can be refined to a composition series.

3. Show that, if $\mathfrak{G}$ has a composition series, then any invariant subgroup of $\mathfrak{G}$ and any factor group of $\mathfrak{G}$ has a composition series. Show also that the composition factors of these series are M-isomorphic to composition factors of $\mathfrak{G}$.

8. The chain conditions. We shall now state two conditions that together are sufficient that an M-group $\mathfrak{G}$ possess a composition series.

I. *Descending chain condition.* If $\mathfrak{G}_1 \supseteq \mathfrak{G}_2 \supseteq \mathfrak{G}_3 \supseteq \cdots$ is a sequence of M-subgroups such that $\mathfrak{G}_1$ is invariant in $\mathfrak{G}$ and each $\mathfrak{G}_{i+1}$ is invariant in the preceding, then there exists a positive integer N such that $\mathfrak{G}_N = \mathfrak{G}_{N+1} = \cdots$.

II. *Ascending chain condition.* If $\mathfrak{H}$ is any term of a normal series and $\mathfrak{H}_1 \subseteq \mathfrak{H}_2 \subseteq \mathfrak{H}_3 \subseteq \cdots$ is an increasing sequence of M-subgroups all of which are invariant in $\mathfrak{H}$, then there exists an integer N such that $\mathfrak{H}_N = \mathfrak{H}_{N+1} = \cdots$.

We remark that, if $\mathfrak{G}$ is commutative, then any subgroup is invariant and any subgroup is a term of a normal series. Hence in this case I and II can be formulated more simply as follows.

III. If $\mathfrak{G}_1 \supseteq \mathfrak{G}_2 \supseteq \mathfrak{G}_3 \supseteq \cdots$ is a descending sequence of M-subgroups, then there exists a positive integer N such that $\mathfrak{G}_N = \mathfrak{G}_{N+1} = \cdots$.

IV. If $\mathfrak{H}_1 \subseteq \mathfrak{H}_2 \subseteq \mathfrak{H}_3 \subseteq \cdots$ is an ascending sequence of M-subgroups, then there exists a positive integer N such that $\mathfrak{H}_N = \mathfrak{H}_{N+1} = \cdots$.

As a matter of fact these conditions can be used also for a non-commutative group if it is known that $\overline{M} = \{\overline{m}\}$ includes all the inner automorphisms of $\mathfrak{G}$; for in this case, too, any M-subgroup is invariant. We shall now prove the following

Theorem 4. *A necessary and sufficient condition that an M-group $\mathfrak{G}$ have a composition series is that $\mathfrak{G}$ satisfies the two chain conditions.*

Sufficiency. We shall show first that if $\mathfrak{H} \neq 1$ is a term of a normal series, then $\mathfrak{H}$ contains a maximal invariant M-subgroup. Thus, either $\mathfrak{H}_1 = 1$ is maximal invariant or there exists a proper invariant M-subgroup $\mathfrak{H}_2$ of $\mathfrak{H}$ such that $\mathfrak{H}_1 \subset \mathfrak{H}_2$. In the latter case if $\mathfrak{H}_2$ is not a maximal invariant M-subgroup of $\mathfrak{H}$, then there is a proper invariant M-subgroup $\mathfrak{H}_3$ of $\mathfrak{H}$ that properly contains $\mathfrak{H}_2$. This process breaks off after a finite number of steps, since otherwise it yields an infinite properly ascending sequence of invariant M-subgroups of $\mathfrak{H}$ contrary to II. Hence, our assertion is proved. In particular we see that $\mathfrak{G} = \mathfrak{G}_1$ contains a maximal invariant M-subgroup $\mathfrak{G}_2$. Also $\mathfrak{G}_2$ contains a maximal invariant M-subgroup $\mathfrak{G}_3$, etc. This gives the properly descending sequence $\mathfrak{G} = \mathfrak{G}_1 \supset \mathfrak{G}_2 \supset \mathfrak{G}_3 \supset \cdots$ in which each $\mathfrak{G}_{i+1}$ is maximal invariant in the preceding. By I there exists a finite number $s + 1$ such that $\mathfrak{G}_{s+1} = 1$.

Necessity. Let $\mathfrak{G}$ have a composition series $\mathfrak{G} = \mathfrak{G}_1 \supset \mathfrak{G}_2 \supset \cdots \supset \mathfrak{G}_{s+1} = 1$ and let $\mathfrak{H}_1 \supset \mathfrak{H}_2 \supset \cdots$ be a properly descending sequence of M-groups such that $\mathfrak{H}_1$ is invariant in $\mathfrak{G}$ and $\mathfrak{H}_{i+1}$ is invariant in $\mathfrak{H}_i$ for $i \geq 1$. Then we assert that the number of $\mathfrak{H}_i$ does not exceed $s + 1$; for, if it does, then $\mathfrak{G} \supseteq \mathfrak{H}_1 \supset \mathfrak{H}_2 \supset \cdots \supset \mathfrak{H}_{s+2} \supseteq 1$ is a normal series. By Schreier's theorem there is a refinement of this series that is equivalent to a refinement of the composition series. If we drop duplicates, we obtain a refinement of the $\mathfrak{H}$-series that is a composition series. But the number of terms exceeds $s + 1$ and this contradicts the Jordan-Hölder theorem. Hence I is proved. A similar argument yields II.

Evidently if $\mathfrak{G}$ is a finite group, then $\mathfrak{G}$ satisfies the chain conditions for any set of operators M. Hence we have composition series for a finite group for any M. A composition series obtained for M vacuous will be called an *ordinary composition series*. Such a series has the form $\mathfrak{G} = \mathfrak{G}_1 \supset \mathfrak{G}_2 \supset \cdots \supset \mathfrak{G}_{s+1} = 1$ where $\mathfrak{G}_{i+1}$ is an invariant subgroup of $\mathfrak{G}_i$ and $\mathfrak{G}_i/\mathfrak{G}_{i+1}$ is a simple group. The Jordan-Hölder theorem proves the invariance of the set of simple groups $\mathfrak{G}_i/\mathfrak{G}_{i+1}$ determined by $\mathfrak{G}$. If $M = \mathfrak{J}$ the set of inner automorphisms, then the M-subgroups are invariant. A composition series in this case has the property that each $\mathfrak{G}_i$ is invariant in $\mathfrak{G}$ and that there exists no invariant subgroup $\mathfrak{G}'$ of $\mathfrak{G}$ such that $\mathfrak{G}_i \supset \mathfrak{G}' \supset \mathfrak{G}_{i+1}$. Such composition series are called *chief series*. Similarly we define a *characteristic series* as a composition series relative to the complete set of automorphisms, and a *fully invariant series* as a composition series relative to the complete set of endomorphisms. The Jordan-Hölder theorem is, of course, applicable to these series, too.

EXERCISES

1. Obtain composition series for S_3 and S_4.
2. Prove that a finite group is solvable if and only if its composition factors are cyclic groups of prime orders.

3. Show that an infinite cyclic group ($M = \varnothing$) satisfies the ascending chain condition but not the descending chain condition.

4. Let $U_{(p)}$ be the multiplicative group of p^k complex roots of unity for p a fixed prime and $k = 0, 1, 2, 3, \cdots$. Show that every proper subgroup of $U_{(p)}$ is finite cyclic. Hence show that $U_{(p)}$ satisfies the descending chain condition but not the ascending chain condition.

9. Direct products. We shall consider in this section a simple construction of an M-group out of n given M-groups $\mathfrak{G}_1$, $\mathfrak{G}_2$, $\cdots$, $\mathfrak{G}_n$. We take $\mathfrak{G}$ to be the product set $\mathfrak{G}_1 \times \mathfrak{G}_2 \times \cdots \times \mathfrak{G}_n$ of elements

$$a = (a_1, a_2, \cdots, a_n), \quad a_i \,\varepsilon\, \mathfrak{G}_i,$$

and we introduce a composition in $\mathfrak{G}$ by the formula

$$(19) \quad (a_1, a_2, \cdots, a_n)(b_1, b_2, \cdots, b_n) = (a_1 b_1, a_2 b_2, \cdots, a_n b_n).$$

If $a = (a_i)$, $b = (b_i)$ and $c = (c_i)$, then

$$(ab)c = ((a_i b_i)c_i) = (a_i(b_i c_i)) = a(bc).$$

Also it is immediate that the element

$$1 = (1, 1, \cdots, 1)$$

is an identity element in $\mathfrak{G}$, and, if we set $a' = (a_i{}^{-1})$, then $aa' = 1 = a'a$. Hence, $\mathfrak{G}$ with our composition is a group. Next we define for $m \,\varepsilon\, M$

$$(20) \qquad (a_1, a_2, \cdots, a_n)m = (a_1 m, a_2 m, \cdots, a_n m).$$

Then

$$(ab)m = ((a_i b_i))m = ((a_i b_i)m) = ((a_i m)(b_i m)) = (am)(bm).$$

Hence our definitions give an M-group. We shall call this M-group the *direct product* of the $\mathfrak{G}_i$ and we use the notation $\mathfrak{G} = \mathfrak{G}_1 \times \mathfrak{G}_2 \times \cdots \times \mathfrak{G}_n$.

It is clear that, if each $\mathfrak{G}_i$ is finite of order n_i, then $\mathfrak{G}$ is finite of order $n = \Pi n_i$. Also $\mathfrak{G}$ is commutative if and only if each $\mathfrak{G}_i$ is commutative. If the additive notation is used in the groups $\mathfrak{G}_i$, it is natural to write

$$(19') \quad (a_1, a_2, \cdots, a_n) + (b_1, b_2, \cdots, b_n)$$

$$= (a_1 + b_1, a_2 + b_2, \cdots, a_n + b_n)$$

in place of (19) and to call $\mathfrak{G}$ the *direct sum* of the $\mathfrak{G}_i$. In this case we write $\mathfrak{G} = \mathfrak{G}_1 \oplus \mathfrak{G}_2 \oplus \cdots \oplus \mathfrak{G}_n$.

The example given in § 1 of the three-dimensional real vector group is precisely the direct sum $\mathfrak{G} \oplus \mathfrak{G} \oplus \mathfrak{G}$ where $\mathfrak{G}$ is the additive group of real numbers relative to the operator set of real numbers and the operation is ordinary multiplication. This is clear from the definitions. The generalization to the n-dimensional vector group is immediate. Another important example of a direct sum is the group $\mathfrak{G} \oplus \mathfrak{G} \oplus \cdots \oplus \mathfrak{G}$ where $\mathfrak{G}$ is the additive group of integers and $M = \varnothing$. The elements of this group are the integral vectors (or "lattice points") with addition the usual vector addition (19′).

We now make two simple remarks about the direct product for arbitrary groups. First, the direct product is independent of the order of the factors. By this we mean that, if $1'$, $2'$, $\cdots$, n' is a permutation of 1, 2, $\cdots$, n, then $\mathfrak{G}_{1'} \times \mathfrak{G}_{2'} \times \cdots \times \mathfrak{G}_{n'}$ is M-isomorphic to $\mathfrak{G}_1 \times \mathfrak{G}_2 \times \cdots \times \mathfrak{G}_n$. In fact it is immediate that the correspondence $(a_1, a_2, \cdots, a_n) \to (a_{1'}, a_{2'}, \cdots, a_{n'})$ is an M-isomorphism. Next we note that, if $n_1 < n_2 < \cdots < n_r = n$, then

$$(\mathfrak{G}_1 \times \cdots \times \mathfrak{G}_{n_1}) \times (\mathfrak{G}_{n_1+1} \times \cdots \times \mathfrak{G}_{n_2}) \times \cdots$$
$$\times (\mathfrak{G}_{n_{r-1}+1} \times \cdots \times \mathfrak{G}_{n_r})$$

is M-isomorphic to $\mathfrak{G}_1 \times \mathfrak{G}_2 \times \cdots \times \mathfrak{G}_n$. Here the mapping

$$(a_1, a_2, \cdots, a_n) \to ((a_1, \cdots, a_{n_1}), \ (a_{n_1+1}, \cdots, a_{n_2}), \ \cdots,$$
$$(a_{n_{r-1}+1}, \cdots, a_{n_r}))$$

is an isomorphism. In particular, it follows that $(\mathfrak{G}_1 \times \mathfrak{G}_2) \times \mathfrak{G}_3$ and $\mathfrak{G}_1 \times (\mathfrak{G}_2 \times \mathfrak{G}_3)$ are equivalent since each is equivalent to $\mathfrak{G}_1 \times \mathfrak{G}_2 \times \mathfrak{G}_3$. Thus, in this sense direct multiplication of groups is associative as well as commutative.

10. Direct products of subgroups. We shall now determine conditions that a given M-group be isomorphic to a direct product. For this purpose we examine further the direct product $\mathfrak{G} = \mathfrak{G}_1 \times \mathfrak{G}_2 \times \cdots \times \mathfrak{G}_n$. Let $\mathfrak{G}_i'$ be the subset of $\mathfrak{G}$ of elements of the form $a_i' = (1, 1, \cdots, 1, a_i, 1, \cdots, 1)$, a_i in the ith position. It is clear that $\mathfrak{G}_i'$ is an M-subgroup of $\mathfrak{G}$ isomorphic to $\mathfrak{G}_i$ under the correspondence

$$a_i \to (1, \cdots, 1, a_i, 1, \cdots, 1).$$

Moreover,

$$(c_1^{-1}, c_2^{-1}, \cdots, c_n^{-1})(1, \cdots, 1, a_i, 1, \cdots, 1)(c_1, c_2, \cdots, c_n)$$

$$= (1, \cdots, 1, c_i^{-1}a_ic_i, 1, \cdots, 1).$$

Hence $\mathfrak{G}_i'$ is invariant in $\mathfrak{G}$. We note next that an arbitrary element $(a_1, a_2, \cdots, a_n)$ of $\mathfrak{G}$ is a product $a_1'a_2'\cdots a_n'$, a_i' in $\mathfrak{G}_i'$. Hence

$$(21) \qquad\qquad \mathfrak{G} = \mathfrak{G}_1'\mathfrak{G}_2' \cdots \mathfrak{G}_n'.$$

In other words, the smallest subgroup of $\mathfrak{G}$ containing all the $\mathfrak{G}_i'$ is $\mathfrak{G}$ itself. Finally, we observe that

$$(22) \quad \mathfrak{G}_i' \cap \mathfrak{G}_1'\mathfrak{G}_2' \cdots \mathfrak{G}_{i-1}'\mathfrak{G}_{i+1}' \cdots \mathfrak{G}_n' = 1, \quad i = 1, 2, \cdots n,$$

since any element in $\mathfrak{G}_1'\mathfrak{G}_2' \cdots \mathfrak{G}_{i-1}'\mathfrak{G}_{i+1}' \cdots \mathfrak{G}_n'$ has the form $(a_1, a_2, \cdots, a_{i-1}, 1, a_{i+1}, \cdots, a_n)$ and any element of $\mathfrak{G}_i'$ has the form $(1, \cdots, 1, a_i, 1, \cdots, 1)$; hence the equality

$$(a_1, a_2, \cdots, a_{i-1}, 1, a_{i+1}, \cdots, a_n) = (1, \cdots, 1, a_i, 1, \cdots, 1)$$

implies that each $a_j = 1$. Thus, any element common to $\mathfrak{G}_i'$ and $\mathfrak{G}_1' \cdots \mathfrak{G}_{i-1}'\mathfrak{G}_{i+1}' \cdots \mathfrak{G}_n'$ has all of its components $a_j = 1$ and this proves (22). We have therefore established the necessity part of the following

Theorem 5. *A necessary and sufficient condition that an M-group $\mathfrak{G}$ be isomorphic to a direct product $\mathfrak{G}_1 \times \mathfrak{G}_2 \times \cdots \times \mathfrak{G}_n$ is that $\mathfrak{G}$ contain invariant M-subgroups $\mathfrak{G}_i'$ isomorphic to $\mathfrak{G}_i$ such that (21) and (22) hold.*

It remains to prove that the condition is sufficient. Hence we suppose that our M-group $\mathfrak{G}$ contains the invariant M-subgroups $\mathfrak{G}_i'$ isomorphic to $\mathfrak{G}_i$ and satisfying (21) and (22). By (21) any element of $\mathfrak{G}$ has the form $a_1'a_2' \cdots a_n'$, a_i' in $\mathfrak{G}_i'$. Let $i \neq j$ and consider the product $a_i'a_j'(a_i')^{-1}(a_j')^{-1}$. Since $a_i'(a_j')(a_i')^{-1} \varepsilon \mathfrak{G}_j'$, $a_i'a_j'(a_i')^{-1}(a_j')^{-1}$ is in $\mathfrak{G}_j'$. Since $a_j'(a_i')^{-1}(a_j')^{-1} \varepsilon \mathfrak{G}_i'$, $a_i'a_j'(a_i')^{-1}(a_j')^{-1} \varepsilon \mathfrak{G}_i'$. Now by (22) $\mathfrak{G}_i' \cap \mathfrak{G}_j' = 1$. Hence

$$a_i'a_j'(a_i')^{-1}(a_j')^{-1} = 1 \quad \text{and} \quad a_i'a_j' = a_j'a_i'.$$

This shows that any element of one of the groups $\mathfrak{G}_i'$ commutes with any element of a different $\mathfrak{G}_j'$. This implies that, if $a_i' \ \varepsilon \ \mathfrak{G}_i'$ and $b_i' \ \varepsilon \ \mathfrak{G}_i'$, then

$$(23) \quad (a_1'a_2' \cdots a_n')(b_1'b_2' \cdots b_n') = (a_1'b_1')(a_2'b_2') \cdots (a_n'b_n').$$

We now consider the direct product $\mathfrak{G}_1 \times \mathfrak{G}_2 \times \cdots \times \mathfrak{G}_n$. Let $a_i \to a_i'$ be an isomorphism of $\mathfrak{G}_i$ onto $\mathfrak{G}_i'$. Then we shall show that the mapping

$$(24) \qquad\qquad (a_1, a_2, \cdots, a_n) \to a_1'a_2' \cdots a_n'$$

is an isomorphism of $\mathfrak{G}_1 \times \mathfrak{G}_2 \times \cdots \times \mathfrak{G}_n$ onto $\mathfrak{G}$. Since

$$(a_1, a_2, \cdots, a_n)(b_1, b_2, \cdots, b_n) = (a_1b_1, a_2b_2, \cdots, a_nb_n) \to$$
$$(a_1b_1)'(a_2b_2)' \cdots (a_nb_n)'$$
$$= (a_1'b_1')(a_2'b_2') \cdots (a_n'b_n')$$
$$= (a_1'a_2' \cdots a_n')(b_1'b_2' \cdots b_n')$$

by (23), the mapping (24) is a homomorphism. Since $(a_1, a_2, \cdots, a_n)m = (a_1m, a_2m, \cdots, a_nm) \to (a_1'm)(a_2'm) \cdots (a_n'm) = (a_1'a_2' \cdots a_n')m$, the mapping is an M-mapping. The mapping is a mapping onto $\mathfrak{G}$ since any element of $\mathfrak{G}$ has the form $a_1'a_2' \cdots a_n'$, a_i' in $\mathfrak{G}_i'$. Finally, we prove that the mapping is an isomorphism by showing that the kernel is the identity. Thus let $a_1'a_2' \cdots a_n' = 1$. Then

$$(a_i')^{-1} = a_1'a_2' \cdots a_{i-1}'a_{i+1}' \cdots a_n',$$

and by (22) $a_i' = 1$. Hence, each $a_i = 1$ and this proves our assertion.

Because of this result we shall say that an M-group $\mathfrak{G}$ is a *direct product of the invariant M-subgroups $\mathfrak{G}_1, \mathfrak{G}_2, \cdots, \mathfrak{G}_n$* if the $\mathfrak{G}_i$ satisfy

$$(25) \quad \mathfrak{G} = \mathfrak{G}_1\mathfrak{G}_2 \cdots \mathfrak{G}_n, \quad \mathfrak{G}_i \cap (\mathfrak{G}_1 \cdots \mathfrak{G}_{i-1}\mathfrak{G}_{i+1} \cdots \mathfrak{G}_n) = 1.$$

Strictly speaking, of course, we can assert only that $\mathfrak{G}$ is isomorphic to the direct product $\mathfrak{G}_1 \times \mathfrak{G}_2 \times \cdots \times \mathfrak{G}_n$. For the sake of simplicity we do not emphasize this distinction and we write $\mathfrak{G} = \mathfrak{G}_1 \times \mathfrak{G}_2 \times \cdots \times \mathfrak{G}_n$.

As an illustration of the criterion given in Theorem 5 we prove now the following

Theorem 6. *If $\mathfrak{G}$ is a finite cyclic group of order $n = p_1^{e_1} p_2^{e_2} \cdots p_s^{e_s}$, p_i prime, $p_i \neq p_j$ if $i \neq j$, then $\mathfrak{G}$ is a direct product of cyclic groups of orders $p_i^{e_i}$, $i = 1, 2, \cdots, s$.*

Proof. Let $\mathfrak{G}_i$ be the subgroup of order $p_i^{e_i}$ and set $\mathfrak{G}' = \mathfrak{G}_1 \mathfrak{G}_2 \cdots \mathfrak{G}_s$. This subgroup has order n' divisible by $p_i^{e_i}$ since $\mathfrak{G}' \supseteq \mathfrak{G}_i$. Hence n' is divisible by $n = p_1^{e_1} p_2^{e_2} \cdots p_s^{e_s}$. It follows that $n' = n$ and that $\mathfrak{G}' = \mathfrak{G}$. Next let $\mathfrak{H}_i$ be the subgroup of $\mathfrak{G}$ of order $n_i = n/p_i^{e_i}$. Let $\mathfrak{Z}_i = \mathfrak{H}_i \cap \mathfrak{G}_i$. Then $\mathfrak{Z}_i$ is a subgroup of $\mathfrak{G}$ whose order is a divisor of n_i and of $p_i^{e_i}$. Since $(n_i, p_i^{e_i}) = 1$, this implies that $\mathfrak{Z}_i = 1$, that is, $\mathfrak{H}_i \cap \mathfrak{G}_i = 1$. Since the order of $\mathfrak{H}_i$ is divisible by $p_j^{e_j}$, $j \neq i$, $\mathfrak{H}_i \supseteq \mathfrak{G}_j$. Hence $\mathfrak{H}_i \supseteq \mathfrak{G}_1 \cdots \mathfrak{G}_{i-1}\mathfrak{G}_{i+1} \cdots \mathfrak{G}_s$. Hence $\mathfrak{G}_1 \cdots \mathfrak{G}_{i-1}\mathfrak{G}_{i+1} \cdots \mathfrak{G}_s \cap \mathfrak{G}_i = 1$ for $i = 1, 2, \cdots, s$ and the conditions of Theorem 5 are fulfilled.

The conditions (21) and (22) of Theorem 5 concern relations among the subgroups $\mathfrak{G}_i$. It is often easier to verify the element conditions given in the following

Theorem 7. *If $\mathfrak{G}$ contains M-subgroups $\mathfrak{G}_i$, $i = 1, 2, \cdots, n$, such that (1) $a_i a_j = a_j a_i$ for any $a_i \,\varepsilon\, \mathfrak{G}_i$ and any $a_j \,\varepsilon\, \mathfrak{G}_j$, $i \neq j$, and (2) every element of $\mathfrak{G}$ can be written in one and only one way as a product $a_1 a_2 \cdots a_n$, a_i in $\mathfrak{G}_i$, then $\mathfrak{G} = \mathfrak{G}_1 \times \mathfrak{G}_2 \times \cdots \times \mathfrak{G}_n$.*

Proof. We note first that each $\mathfrak{G}_i$ is invariant in $\mathfrak{G}$; for, if $g_i \,\varepsilon\, \mathfrak{G}_i$ and $a = a_1 a_2 \cdots a_n$, $a_j \,\varepsilon\, \mathfrak{G}_j$, then

$$a^{-1}g_i a = a_n^{-1} \cdots a_2^{-1} a_1^{-1} g_i a_1 a_2 \cdots a_n = a_i^{-1} g_i a_i \,\varepsilon\, \mathfrak{G}_i$$

by (1). Since by (2), a can represent any element of $\mathfrak{G}$, $\mathfrak{G}_i$ is invariant in $\mathfrak{G}$. Also by (2) $\mathfrak{G} = \mathfrak{G}_1 \mathfrak{G}_2 \cdots \mathfrak{G}_n$. Any element of $\mathfrak{G}_1 \cdots \mathfrak{G}_{i-1}\mathfrak{G}_{i+1} \cdots \mathfrak{G}_n$ has the form $a_1 a_2 \cdots a_{i-1}a_{i+1} \cdots a_n$, a_j in $\mathfrak{G}_j$. If this element is also in $\mathfrak{G}_i$, then we have

$$a_i = a_1 a_2 \cdots a_{i-1}a_{i+1} \cdots a_n, \quad a_j \text{ in } \mathfrak{G}_j.$$

Hence

$$1 \cdots 1 a_i 1 \cdots 1 = a_1 a_2 \cdots a_{i-1} 1 a_{i+1} \cdots a_n.$$

Since there is only one way of writing an element as a product $a_1 a_2 \cdots a_n$, a_j in $\mathfrak{G}_j$, this gives $a_i = 1$. Hence $\mathfrak{G}_i \cap \mathfrak{G}_1 \cdots \mathfrak{G}_{i-1}\mathfrak{G}_{i+1} \cdots \mathfrak{G}_n = 1$. The theorem now follows from our first criterion.

We remark also that the conditions (21) and (22) imply the conditions (1) and (2) of the present theorem. This was established in the proof of Theorem 5.

The following important results on direct products of subgroups can now be easily derived:

A. If $\mathfrak{G} = \mathfrak{G}_1 \times \mathfrak{G}_2 \times \cdots \times \mathfrak{G}_n$, then $\mathfrak{G} = \mathfrak{H}_1 \times \mathfrak{H}_2 \times \cdots \times \mathfrak{H}_r$ where $\mathfrak{H}_1 = \mathfrak{G}_1\mathfrak{G}_2 \cdots \mathfrak{G}_{n_1}$, $\mathfrak{H}_2 = \mathfrak{G}_{n_1+1}\mathfrak{G}_{n_1+2} \cdots \mathfrak{G}_{n_2}$, $\cdots$, $\mathfrak{H}_r = \mathfrak{G}_{n_{r-1}+1}\mathfrak{G}_{n_{r-1}+2} \cdots \mathfrak{G}_{n_r}$. Also

$$
\begin{aligned}
\mathfrak{H}_1 &= \mathfrak{G}_1 \times \mathfrak{G}_2 \times \cdots \times \mathfrak{G}_{n_1}, \\
(26) \qquad \mathfrak{H}_2 &= \mathfrak{G}_{n_1+1} \times \mathfrak{G}_{n_1+2} \times \cdots \times \mathfrak{G}_{n_2}, \\
&\;\; \cdots \cdots \cdots \cdots \cdots \cdots \\
\mathfrak{H}_r &= \mathfrak{G}_{n_{r-1}+1} \times \mathfrak{G}_{n_{r-1}+2} \times \cdots \times \mathfrak{G}_{n_r}.
\end{aligned}
$$

B. If $\mathfrak{G} = \mathfrak{H}_1 \times \mathfrak{H}_2 \times \cdots \times \mathfrak{H}_r$ and (26) holds, then $\mathfrak{G} = \mathfrak{G}_1 \times \mathfrak{G}_2 \times \cdots \times \mathfrak{G}_n$.

We omit the proofs. We note also the following result.

C. If $\mathfrak{G} = \mathfrak{G}_1 \times \mathfrak{G}_2$, then $\mathfrak{G}_2 \cong \mathfrak{G}/\mathfrak{G}_1$. This follows directly from the second isomorphism theorem; for $\mathfrak{G}_1$ is invariant in $\mathfrak{G}$. Hence, $\mathfrak{G}/\mathfrak{G}_1 = \mathfrak{G}_1\mathfrak{G}_2/\mathfrak{G}_1 \cong \mathfrak{G}_2/\mathfrak{G}_1 \cap \mathfrak{G}_2 = \mathfrak{G}_2/1 \cong \mathfrak{G}_2$.

EXERCISES

1. Prove Theorem 6 by showing directly that, if b is an element of order $n = p_1^{e_1}p_2^{e_2} \cdots p_s^{e_s}$, then $b = b_1 b_2 \cdots b_s$ where b_i has order $p_i^{e_i}$.

2. Prove that, if $\mathfrak{G}$ is cyclic of order $n = st$, $(s,t) = 1$, then $\mathfrak{G} = \mathfrak{H} \times \mathfrak{K}$ where $\mathfrak{H}$ is of order s and $\mathfrak{K}$ is of order t.

3. Prove that, if $\mathfrak{G}$ is a finite commutative group of order $n = p_1^{e_1}p_2^{e_2} \cdots p_s^{e_s}$, p_i distinct primes, then $\mathfrak{G} = \mathfrak{G}_1 \times \mathfrak{G}_2 \times \cdots \times \mathfrak{G}_s$ where $\mathfrak{G}_i$ is a subgroup all of whose elements have order a power of p_i.

11. Projections. Let $\mathfrak{G} = \mathfrak{G}_1 \times \mathfrak{G}_2 \times \cdots \times \mathfrak{G}_n$ where the $\mathfrak{G}_i$ are subgroups, and let η_i, $i = 1, 2, \cdots, n$, be a homomorphism of $\mathfrak{G}_i$ into another M-group $\bar{\mathfrak{G}}$. Assume, moreover, that, if $x_i \, \varepsilon \, \mathfrak{G}_i$, $x_j \, \varepsilon \, \mathfrak{G}_j$ and $i \neq j$, then $(x_i\eta_i)(x_j\eta_j) = (x_j\eta_j)(x_i\eta_i)$. Now we can write any $x \, \varepsilon \, \mathfrak{G}$ as $x_1 x_2 \cdots x_n$, x_i in $\mathfrak{G}_i$, and we can define a mapping η of $\mathfrak{G}$ into $\bar{\mathfrak{G}}$ by

$$(27) \qquad (x_1 x_2 \cdots x_n)\eta = (x_1 \eta_1)(x_2 \eta_2) \cdots (x_n \eta_n).$$

We can verify directly that η is an M-homomorphism of $\mathfrak{G}$ into $\bar{\mathfrak{G}}$.

This method of putting together M-homomorphisms of the $\mathfrak{G}_i$ is particularly important if $\bar{\mathfrak{G}}$, too, is a direct product. Thus let $\bar{\mathfrak{G}} = \bar{\mathfrak{G}}_1 \times \bar{\mathfrak{G}}_2 \times \cdots \times \bar{\mathfrak{G}}_n$ and let η_i be a homomorphism of $\mathfrak{G}_i$ into $\bar{\mathfrak{G}}_i$. Then $x_i \eta_i \ \varepsilon \ \bar{\mathfrak{G}}_i$ and $x_j \eta_j \ \varepsilon \ \bar{\mathfrak{G}}_j$; hence, if $i \neq j$, $(x_i \eta_i)(x_j \eta_j) = (x_j \eta_j)(x_i \eta_i)$. It follows that the mapping given by (27) is an M-homomorphism of $\mathfrak{G}$ into $\bar{\mathfrak{G}}$.

We apply this remark first to define certain endomorphisms that can be associated with a direct decomposition of $\mathfrak{G}$ as $\mathfrak{G}_1 \times \mathfrak{G}_2 \times \cdots \times \mathfrak{G}_n$. We define ϵ_i to be the endomorphism of $\mathfrak{G}$ that is obtained by putting together in the manner indicated the endomorphisms

$$x_1 \to 1, \ \cdots, \ x_{i-1} \to 1, \ x_i \to x_i, \ x_{i+1} \to 1, \ \cdots, \ x_n \to 1.$$

Then by (27)

$$(28) \qquad x\epsilon_i = (x_1 x_2 \cdots x_n)\epsilon_i = x_i.$$

If x_i is any element of $\mathfrak{G}_i$, the decomposition of x_i as a product of elements of the $\mathfrak{G}_j$ reads $x_i = 1 \cdots 1 x_i 1 \cdots 1$. Hence, it is clear from (28) that $x_i \epsilon_i = x_i$ and $x_i \epsilon_j = 1$ if $i \neq j$. If x is any element of $\mathfrak{G}$, then $x\epsilon_i = x_i \ \varepsilon \ \mathfrak{G}_i$. Hence, $(x\epsilon_i)\epsilon_i = x\epsilon_i$ and $(x\epsilon_i)\epsilon_j = 1$. Thus, if we denote the endomorphism $x \to 1$ by 0, then we have proved that

$$(29) \qquad \epsilon_i{}^2 = \epsilon_i, \quad \epsilon_i \epsilon_j = 0 \quad \text{if} \quad i \neq j.$$

We note next that the mappings ϵ_i are *normal* in the sense that they commute with all the inner automorphisms of $\mathfrak{G}$; for, if $x = x_1 x_2 \cdots x_n$ where $x_i \ \varepsilon \ \mathfrak{G}_i$ and a is any other element of $\mathfrak{G}$, then

$$a^{-1}xa = (a^{-1}x_1 a)(a^{-1}x_2 a) \cdots (a^{-1}x_n a)$$

and $a^{-1}x_i a \ \varepsilon \ \mathfrak{G}_i$. Hence

$$(a^{-1}xa)\epsilon_i = a^{-1}x_i a = a^{-1}(x\epsilon_i)a$$

and this proves our assertion. Now we shall call an M-endomorphism ϵ a *projection* if ϵ is normal and idempotent ($\epsilon^2 = \epsilon$). A pair of projections ϵ, ϵ' will be called *orthogonal* if $\epsilon\epsilon' = 0 = \epsilon'\epsilon$.

Using these terms, we can say that the ϵ_i determined by the decomposition $\mathfrak{G} = \mathfrak{G}_1 \times \mathfrak{G}_2 \times \cdots \times \mathfrak{G}_n$ are orthogonal projections.

There is another important relation connecting the ϵ_i. This involves a second important composition of mappings in a group. If η_1 and η_2 are two mappings of the group $\mathfrak{G}$ into itself, then we define the *sum* $\eta_1 + \eta_2$ by

$$(30) \qquad x(\eta_1 + \eta_2) = (x\eta_1)(x\eta_2).$$

We have considered this composition before in the case of endomorphisms of a commutative group (§ **12**, Chapter II). We have seen that it, together with the product as resultant, turns the set of endomorphisms of a commutative group into a ring. In the non-commutative case the sum of two endomorphisms need not be an endomorphism.

It is immediate from (30) that the sum composition for arbitrary mappings of $\mathfrak{G}$ into itself is associative but not necessarily commutative. The endomorphism 0 $(x \to 1)$ acts as an identity for addition since

$$x(\eta + 0) = (x\eta)(x0) = (x\eta)1 = x\eta$$

$$x(0 + \eta) = (x0)(x\eta) = 1(x\eta) = x\eta.$$

Also, if we define $-\eta$ by $x(-\eta) = (x\eta)^{-1}$, then

$$x(-\eta + \eta) = (x\eta)^{-1}(x\eta) = 1$$

$$x(\eta + (-\eta)) = (x\eta)(x\eta)^{-1} = 1.$$

Hence $-\eta + \eta = 0 = \eta + (-\eta)$. This proves that the set of mappings of $\mathfrak{G}$ together with the addition composition is a group.

Multiplication of mappings is right distributive relative to addition:

$$(31) \qquad \rho(\eta_1 + \eta_2) = \rho\eta_1 + \rho\eta_2;$$

since

$$x\rho(\eta_1 + \eta_2) = ((x\rho)\eta_1)((x\rho)\eta_2),$$

$$x(\rho\eta_1 + \rho\eta_2) = (x(\rho\eta_1))(x(\rho\eta_2)) = ((x\rho)\eta_1)((x\rho)\eta_2).$$

The other distributive law does not hold in general. However, it is valid if ρ is an endomorphism, since

$$x((\eta_1 + \eta_2)\rho) = ((x\eta_1)(x\eta_2))\rho = ((x\eta_1)\rho)((x\eta_2)\rho)$$

$$= (x(\eta_1\rho))(x(\eta_2\rho)) = x(\eta_1\rho + \eta_2\rho).$$

We return now to our investigation of the projections ϵ_i determined by the direct decomposition $\mathfrak{G} = \mathfrak{G}_1 \times \mathfrak{G}_2 \times \cdots \times \mathfrak{G}_n$. If x is any element of $\mathfrak{G}$, $x = x_1 x_2 \cdots x_n$, x_i in $\mathfrak{G}_i$. Hence $x = (x\epsilon_1)(x\epsilon_2) \cdots (x\epsilon_n)$ so that by the definitions of addition and of 1,

$$(32) \qquad \epsilon_1 + \epsilon_2 + \cdots + \epsilon_n = 1.$$

The properties (29) and (32) are characteristic of the projections determined by a direct decomposition. Thus suppose that $\epsilon_1, \epsilon_2, \cdots, \epsilon_n$ are normal M-endomorphisms satisfying (29) and (32). Then $\mathfrak{G}_i = \mathfrak{G}\epsilon_i$ is an M-subgroup and $\mathfrak{G}_i$ is invariant, since

$$a^{-1}(x\epsilon_i)a = (a^{-1}xa)\epsilon_i$$

is in $\mathfrak{G}_i$. Since $x = x1 = x(\epsilon_1 + \epsilon_2 + \cdots + \epsilon_n) = (x\epsilon_1)(x\epsilon_2) \cdots (x\epsilon_n)$, $\mathfrak{G} = \mathfrak{G}_1 \mathfrak{G}_2 \cdots \mathfrak{G}_n$. We note next that since $\mathfrak{G}_i = \mathfrak{G}\epsilon_i$, ϵ_i is the identity mapping in $\mathfrak{G}_i$. Also if $j \neq i$, then ϵ_i maps $\mathfrak{G}_j$ into 1. Hence if $z \in \mathfrak{G}_i \cap \mathfrak{G}_1 \mathfrak{G}_2 \cdots \mathfrak{G}_{i-1} \mathfrak{G}_{i+1} \cdots \mathfrak{G}_n$, $z\epsilon_i = z$ and $z\epsilon_i = 1$. Hence

$$\mathfrak{G}_i \cap \mathfrak{G}_1 \mathfrak{G}_2 \cdots \mathfrak{G}_{i-1} \mathfrak{G}_{i+1} \cdots \mathfrak{G}_n = 1,$$

and $\mathfrak{G} = \mathfrak{G}_1 \times \mathfrak{G}_2 \times \cdots \times \mathfrak{G}_n$. Since $x = (x\epsilon_1)(x\epsilon_2) \cdots (x\epsilon_n)$, $x\epsilon_i$ in $\mathfrak{G}_i$, the projections determined by this decomposition are the given mappings ϵ_i. This closes the circle in our considerations.

EXERCISES

1. Show that if η is a normal endomorphism, then η has the form $a\eta = c(a,\eta)a$ where $c(a,\eta)$ is an element that commutes with every element of $\mathfrak{G}\eta$ and $c(ab,\eta) = c(a,\eta)[ac(b,\eta)a^{-1}]$.

2. Prove that, if the center $\mathfrak{C} = 1$ or if the commutator group $\mathfrak{G}^{(1)} = \mathfrak{G}$ (definition in ex. 3, p. 132), then the identity mapping is the only normal automorphism of $\mathfrak{G}$.

3. Let $\epsilon_1, \epsilon_2, \cdots, \epsilon_n$ be the projections of a direct decomposition. Show that, if $i_1, i_2, \cdots, i_r$ are distinct, then $\epsilon_{i_1} + \epsilon_{i_2} + \cdots + \epsilon_{i_r}$ is an endomorphism. Show also that $\epsilon_i + \epsilon_j = \epsilon_j + \epsilon_i$.

12. Decomposition into indecomposable groups.

An M-group $\mathfrak{G}$ is said to be *decomposable* if $\mathfrak{G} = \mathfrak{G}_1 \times \mathfrak{G}_2$ where each $\mathfrak{G}_i$ is a proper subgroup. Then also $\mathfrak{G}_i \neq 1$. Hence the projection ϵ_i, $i = 1,2$, is $\neq 1$, $\neq 0$. Thus, if $\mathfrak{G}$ is decomposable, then there exist projections of $\mathfrak{G}$ that are $\neq 1,0$. Conversely, this condition is sufficient for $\mathfrak{G}$ to be decomposable; for let ϵ_1 be a projection $\neq 1,0$. Put $\mathfrak{G}_1 = \mathfrak{G}\epsilon_1$ and let $\mathfrak{G}_2$ be the kernel of the endomorphism ϵ_1. Then $\mathfrak{G}_1$ and $\mathfrak{G}_2$ are M-subgroups and, because of the normality of ϵ_1, both of these subgroups are invariant. If x is any element of $\mathfrak{G}$, $z = x(-\epsilon_1 + 1) = (x\epsilon_1)^{-1}x$ is in $\mathfrak{G}_2$, since

$$((x\epsilon_1)^{-1}x)\epsilon_1 = ((x\epsilon_1)^{-1}\epsilon_1)(x\epsilon_1) = (x\epsilon_1{}^2)^{-1}(x\epsilon_1) = 1.$$

Hence, $x = (x\epsilon_1)z \; \varepsilon \; \mathfrak{G}_1\mathfrak{G}_2$. Also, if x_1 is any element of $\mathfrak{G}_1$, then $x_1 = x\epsilon_1$ for a suitable x in $\mathfrak{G}$. Hence $x_1 = x\epsilon_1 = x\epsilon_1{}^2 = x_1\epsilon_1$. Hence $\mathfrak{G}_1 \cap \mathfrak{G}_2 = 1$. Thus $\mathfrak{G} = \mathfrak{G}_1 \times \mathfrak{G}_2$. Since $\epsilon_1 \neq 1,0$, $\mathfrak{G}_1 \neq \mathfrak{G}$ and $\mathfrak{G}_2 \neq \mathfrak{G}$ and $\mathfrak{G}$ is decomposable. We can therefore state the following

Theorem 8. *A necessary and sufficient condition that an M-group be decomposable is that there exist projections of $\mathfrak{G}$ that are $\neq 1$, $\neq 0$.*

We show next that any group $\mathfrak{G} \neq 1$ satisfying the descending chain condition for invariant M-subgroups permits a decomposition into indecomposable M-groups. The assumption we are making is

I'. If $\mathfrak{G}_1 \supseteq \mathfrak{G}_2 \supseteq \mathfrak{G}_3 \supseteq \cdots$ is a decreasing sequence of invariant M-subgroups of $\mathfrak{G}$, then there exists an integer N such that $\mathfrak{G}_N = \mathfrak{G}_{N+1} = \cdots$.

We use this condition to show first that $\mathfrak{G}$ has an indecomposable direct factor; for either $\mathfrak{G}$ is indecomposable or $\mathfrak{G} = \mathfrak{G}_1 \times \mathfrak{G}_2$, where $\mathfrak{G}_1 \neq \mathfrak{G}$, $\neq 1$. If $\mathfrak{G}_1$ is indecomposable, we have the desired factor. Otherwise, $\mathfrak{G}_1 = \mathfrak{G}_{11} \times \mathfrak{G}_{12}$ where $\mathfrak{G}_{11} \neq \mathfrak{G}_1$, 1. Then $\mathfrak{G} \supset \mathfrak{G}_1 \supset \mathfrak{G}_{11}$ and either $\mathfrak{G}_{11}$ is indecomposable or $\mathfrak{G}_{11} = \mathfrak{G}_{111} \times \mathfrak{G}_{112}$ with $\mathfrak{G}_{111} \neq \mathfrak{G}_{11}$, 1. This gives the larger chain $\mathfrak{G} \supset \mathfrak{G}_1 \supset \mathfrak{G}_{11} \supset \mathfrak{G}_{111}$. All of the groups thus obtained are invariant M-subgroups of $\mathfrak{G}$. Hence I' guarantees that this process leads in a finite number of steps to an indecomposable direct factor.

Now let $\mathfrak{G}_1$ denote an indecomposable direct factor of $\mathfrak{G}$ and write $\mathfrak{G} = \mathfrak{G}_1 \times \mathfrak{G}_1{}'$. If $\mathfrak{G}_1{}' \neq 1$, we can factor $\mathfrak{G}_1{}' = \mathfrak{G}_2 \times \mathfrak{G}_2{}'$

where $\mathfrak{G}_2$ is indecomposable. Then $\mathfrak{G} = \mathfrak{G}_1 \times \mathfrak{G}_2 \times \mathfrak{G}_2'$ and $\mathfrak{G}_2'$ is invariant in $\mathfrak{G}$. Next either $\mathfrak{G}_2' = 1$ or $\mathfrak{G}_2' = \mathfrak{G}_3 \times \mathfrak{G}_3'$ where $\mathfrak{G}_3$ is indecomposable. As before $\mathfrak{G}_3'$ is invariant in $\mathfrak{G}$. This process leads to a properly descending chain of invariant M-subgroups $\mathfrak{G} \supset \mathfrak{G}_1' \supset \mathfrak{G}_2' \supset \mathfrak{G}_3' \supset \cdots$. By a second application of I' we conclude that $\mathfrak{G}_n' = 1$ for some integer n. Hence $\mathfrak{G} = \mathfrak{G}_1 \times \mathfrak{G}_2 \times \cdots \times \mathfrak{G}_n$ where the $\mathfrak{G}_i$ are indecomposable. This proves

Theorem 9. *Any M-group $\neq 1$ that satisfies the descending chain condition for invariant M-subgroups can be expressed as a direct product of a finite number of indecomposable groups $\neq 1$.*

13. The Krull-Schmidt theorem. In this section we shall prove a uniqueness theorem for direct decompositions into indecomposable groups. In order to establish this result we require in addition to the descending chain condition I' the following ascending chain condition:

II′. If $\mathfrak{G}_1 \subseteq \mathfrak{G}_2 \subseteq \mathfrak{G}_3 \subseteq \cdots$ is an ascending sequence of invariant M-subgroups, there exists an N such that $\mathfrak{G}_N = \mathfrak{G}_{N+1} = \cdots$.

We consider first some important consequences of the chain conditions. We prove first the following

Theorem 10. *Let $\mathfrak{G}$ be an M-group that satisfies the descending and the ascending chain conditions for invariant M-subgroups. Then if η is a normal M-endomorphism, η is an automorphism if either (1) η is 1–1 or (2) $\mathfrak{G}\eta = \mathfrak{G}$.*

Proof. Assume that η is 1–1. Then if $\mathfrak{G}\eta^{r-1} = \mathfrak{G}\eta^r$ for some $r = 1, 2, \cdots$, any $y \, \varepsilon \, \mathfrak{G}\eta^{r-2}$ has the property that $y\eta = x\eta^r = (x\eta^{r-1})\eta$ for a suitable element x. Hence $y = x\eta^{r-1} \, \varepsilon \, \mathfrak{G}\eta^{r-1}$. Thus also $\mathfrak{G}\eta^{r-2} = \mathfrak{G}\eta^{r-1}$. If we repeat the argument and continue in this way, we obtain finally $\mathfrak{G} = \mathfrak{G}\eta$. We therefore see that, if $\mathfrak{G} \supset \mathfrak{G}\eta$, then $\mathfrak{G} \supset \mathfrak{G}\eta \supset \mathfrak{G}\eta^2 \supset \cdots$ is an infinite properly descending chain. Since η is a normal M-endomorphism, all the terms of this chain are invariant M-subgroups. We therefore have a contradiction to I'. Hence if η is 1–1, $\mathfrak{G} = \mathfrak{G}\eta$ and so η is an automorphism. Assume next that $\mathfrak{G} = \mathfrak{G}\eta$. Let

3_k denote the kernel of the endomorphism η^k, $k = 0, 1, 2, \cdots$, $\eta^0 \equiv 1$. Since we have adopted the convention that $\eta^0 = 1$, $3_0 = 1$. Also it is clear that $3_{k-1} \subseteq 3_k$. Suppose now that $3_{r-1} = 3_r$ and let $z \,\varepsilon\, 3_{r-1}$. We can write $z = y\eta$. Then $1 = z\eta^{r-1} = (y\eta)\eta^{r-1} = y\eta^r$. Hence $y\eta^{r-1} = 1$, and $z = y\eta$ is sent into 1 by η^{r-2}. Thus $z \,\varepsilon\, 3_{r-2}$. This shows that $3_{r-2} = 3_{r-1}$ and continuing in this way we see that all the $3_k = 1$. Hence, either $3_1 = 1$ or $1 = 3_0 \subset 3_1 \subset 3_2 \subset \cdots$ is an infinite properly ascending chain of invariant M-subgroups. This contradicts II'. Hence we see that, if $\mathfrak{G}\eta = \mathfrak{G}$, then $3_1 = 1$ and η is 1–1.

If η is any endomorphism of a group, we call the totality of elements z such that $z\eta^s = 1$ for some integer s, the *radical* of η. Thus the radical $\mathfrak{R}$ is the set-theoretic sum of the kernels 3_i of the homomorphisms η^i. We use this concept to state the following theorem which is the crucial step in the proof of the uniqueness theorem.

Theorem 11 (Fitting's lemma). *Let $\mathfrak{G}$ be an M-group that satisfies the chain conditions for invariant M-subgroups and let η be a normal M-endomorphism of $\mathfrak{G}$. Then $\mathfrak{G} = \mathfrak{R} \times \mathfrak{H}$ where $\mathfrak{R}$ is the radical of η and $\mathfrak{H}$ satisfies the condition $\mathfrak{H}\eta = \mathfrak{H}$.*

Proof. We have the descending chain of invariant M-subgroups $\mathfrak{G} \supseteq \mathfrak{G}\eta \supseteq \mathfrak{G}\eta^2 \supseteq \cdots$. Hence there is an integer r such that $\mathfrak{G}\eta^r = \mathfrak{G}\eta^{r+1}$. Then $\mathfrak{G}\eta^r = \mathfrak{G}\eta^{r+1} = \mathfrak{G}\eta^{r+2} = \cdots$. Let $\mathfrak{H}$ denote this invariant M-subgroup. Next consider the ascending chain $3_0 \subseteq 3_1 \subseteq 3_2 \subseteq \cdots$ where 3_i is the kernel of η^i. Then there is an integer s such that $3_s = 3_{s+1}$. It follows directly that $3_{s+1} = 3_{s+2} = \cdots$. Hence 3_s is the radical $\mathfrak{R}$ of η. Let t be the larger of the two integers, r,s. If x is any element in $\mathfrak{G}$, $x\eta^t = y\eta^{2t}$ for a suitable y. Hence $x = [x(y\eta^t)^{-1}](y\eta^t)$ and $[x(y\eta^t)^{-1}]\eta^t = (x\eta^t)(y\eta^{2t})^{-1} = 1$. Thus, if we set $z = x(y\eta^t)^{-1}$, then $z\eta^t = 1$ and $z \,\varepsilon\, \mathfrak{R}$. Since $y\eta^t \,\varepsilon\, \mathfrak{H}$ we have the decomposition $\mathfrak{G} = \mathfrak{R}\mathfrak{H}$. Now let $w \,\varepsilon\, \mathfrak{R} \cap \mathfrak{H}$. Then $w = u\eta^t$ and $1 = w\eta^t = u\eta^{2t}$. Hence, $u \,\varepsilon\, \mathfrak{R}$ and $u\eta^t = 1$. Thus $w = 1$. Hence $\mathfrak{G} = \mathfrak{R} \times \mathfrak{H}$.

Since $\mathfrak{R} = 3_s$, it is clear that $z\eta^s = 1$ for every $z \,\varepsilon\, \mathfrak{R}$. This means that η is a nilpotent endomorphism in $\mathfrak{R}$. If $\mathfrak{G}$ is indecomposable, either $\mathfrak{G} = \mathfrak{R}$ or $\mathfrak{G} = \mathfrak{H}$. In the first case η is nilpotent and in the second case η is onto so that by Theorem 10 η is an automorphism. This proves

Corollary 1. *If $\mathfrak{G}$ is an indecomposable M-group that satisfies the chain conditions for invariant M-subgroups, then any normal M-endomorphism of $\mathfrak{G}$ is either nilpotent or an automorphism.*

This corollary enables us to prove a very interesting closure property for the normal nilpotent endomorphisms of an indecomposable group, namely,

Corollary 2. *Let $\mathfrak{G}$ be as in Corollary 1 and let η_1 and η_2 be normal nilpotent M-endomorphisms, then, if $\eta_1 + \eta_2$ is an endomorphism, $\eta_1 + \eta_2$ is nilpotent.*

Proof. According to Corollary 1, if $\eta = \eta_1 + \eta_2$ is not nilpotent, then it is an automorphism. Let η^{-1} be its inverse. Evidently this mapping is a normal M-endomorphism and we have $\eta_1\eta^{-1} + \eta_2\eta^{-1} = 1$, or $\lambda_1 + \lambda_2 = 1$ where $\lambda_i = \eta_i\eta^{-1}$. Since η_i is not an automorphism, its kernel is $\neq 1$. Hence this holds for λ_i, too. Hence λ_i is nilpotent. We note next that $\lambda_1 = \lambda_1(\lambda_1 + \lambda_2)$ $= \lambda_1^2 + \lambda_1\lambda_2$ and $\lambda_1 = (\lambda_1 + \lambda_2)\lambda_1 = \lambda_1^2 + \lambda_2\lambda_1$. Hence $\lambda_1\lambda_2 = \lambda_2\lambda_1$ and consequently for any positive integer m

$$(33) \quad (\lambda_1 + \lambda_2)^m$$

$$= \lambda_1^m + \binom{m}{1}\lambda_1^{m-1}\lambda_2 + \binom{m}{2}\lambda_1^{m-2}\lambda_2^2 + \cdots + \lambda_2^m.$$

Now let $\lambda_1^r = 0$, $\lambda_2^s = 0$ and take $m = r + s - 1$ in this relation. This gives the contradiction $1 = 0$.

EXERCISE

1. Let $\mathfrak{G}$ satisfy I$'$ and II$'$ and let η be a normal endomorphism. Let r be the first integer such that $\mathfrak{G}\eta^r = \mathfrak{G}\eta^{r+1}$ and let s be the first integer such that $\mathfrak{Z}_s = \mathfrak{Z}_{s+1}, \mathfrak{Z}_i$ the kernel of η^i. Prove that $r = s$.

We can now prove the main theorem.

The Krull-Schmidt theorem. *Let $\mathfrak{G}$ be an M-group that satisfies the chain conditions for invariant M-subgroups and let*

$$(34) \qquad \mathfrak{G} = \mathfrak{G}_1 \times \mathfrak{G}_2 \times \cdots \times \mathfrak{G}_s,$$

$$(35) \qquad \mathfrak{G} = \mathfrak{H}_1 \times \mathfrak{H}_2 \times \cdots \times \mathfrak{H}_t$$

be two direct decompositions of $\mathfrak{G}$ into indecomposable groups. Then $s = t$ and for a suitable ordering of the $\mathfrak{H}_i$ we have $\mathfrak{H}_i \cong \mathfrak{G}_i$ and

$$(36) \qquad \mathfrak{G} = \mathfrak{H}_1 \times \cdots \times \mathfrak{H}_k \times \mathfrak{G}_{k+1} \times \cdots \times \mathfrak{G}_s,$$

$$k = 1, 2, \cdots, s.$$

Proof. Suppose that we have already obtained a pairing of $\mathfrak{H}_1, \mathfrak{H}_2, \cdots, \mathfrak{H}_{r-1}$ respectively with $\mathfrak{G}_1, \mathfrak{G}_2, \cdots, \mathfrak{G}_{r-1}$ in such a way that $\mathfrak{G}_i \cong \mathfrak{H}_i$, $i = 1, 2, \cdots, r - 1$, and (36) holds for $k \le r - 1$. (At the start we have $r = 1$.) Consider the intermediate decomposition

$$(37) \qquad \mathfrak{G} = \mathfrak{H}_1 \times \mathfrak{H}_2 \times \cdots \times \mathfrak{H}_{r-1} \times \mathfrak{G}_r \times \cdots \times \mathfrak{G}_s.$$

Let $\lambda_1, \lambda_2, \cdots, \lambda_s$ be the projections determined by this decomposition and let $\eta_1, \eta_2, \cdots, \eta_t$ be the projections determined by (35). Evidently we have $\lambda_r = \left(\sum_1^t \eta_j \right) \lambda_r = \sum_1^t \eta_j \lambda_r$. For any x in $\mathfrak{G}$, $x\eta_j \,\varepsilon\, \mathfrak{H}_j$; hence if $j \le r - 1$ we have by (37), $x\eta_j = x\eta_j\lambda_j$ and $x\eta_j\lambda_r = x\eta_j\lambda_j\lambda_r = 1$. Thus $\eta_j\lambda_r = 0$, and we have the relation

$$(38) \qquad \lambda_r = \eta_r\lambda_r + \eta_{r+1}\lambda_r + \cdots + \eta_t\lambda_r.$$

We operate now in $\mathfrak{G}_r$. Here $\lambda_r = 1$ so that $1 = \sum_r^t \eta_j\lambda_r$. Also any partial sum $\Sigma\eta_{i_l}\lambda_r = (\Sigma\eta_{i_l})\lambda_r$ induces a normal M-endomorphism in $\mathfrak{G}_r$. Since $\mathfrak{G}_r$ is indecomposable it follows from Corollary 2 that there exists a u, $r \le u \le t$ such that $\eta_u\lambda_r$ defines an automorphism of $\mathfrak{G}_r$. We can renumber the $\mathfrak{H}_i$, $i = r, r + 1, \cdots$, so that $\mathfrak{H}_u$ becomes $\mathfrak{H}_r$. We proceed to show that $\mathfrak{G}_r \cong \mathfrak{H}_r$ and that (36) holds for $k = r$.

Since $\eta_r\lambda_r$ is an automorphism in $\mathfrak{G}_r$, its kernel is 1. Hence $z\eta_r = 1$ for z in $\mathfrak{G}_r$ implies that $z = 1$. Thus η_r maps $\mathfrak{G}_r$ isomorphically into $\mathfrak{H}_r$. Let $\bar{\mathfrak{H}}_r = \mathfrak{G}_r\eta_r$ and let $\mathfrak{U}_r$ be the subset of $\mathfrak{H}_r$ of elements u such that $u\lambda_r = 1$. Since λ_r is an isomorphism of $\bar{\mathfrak{H}}_r = \mathfrak{G}_r\eta_r$, $\bar{\mathfrak{H}}_r \cap \mathfrak{U}_r = 1$. Also if y is any element of $\mathfrak{H}_r$, then $y\lambda_r \,\varepsilon\, \mathfrak{G}_r$ so that $y\lambda_r = v\eta_r\lambda_r$ for a suitable v in $\mathfrak{G}_r$. We can write $y = (y(v\eta_r)^{-1})(v\eta_r)$ and note that $y(v\eta_r)^{-1} \,\varepsilon\, \mathfrak{U}_r$ and $v\eta_r \,\varepsilon\, \bar{\mathfrak{H}}_r$. Hence $\mathfrak{H}_r = \mathfrak{U}_r\bar{\mathfrak{H}}_r = \mathfrak{U}_r \times \bar{\mathfrak{H}}_r$. Since $\mathfrak{H}_r$ is indecomposable and

$\mathfrak{H}_r \neq 1$, $\mathfrak{H}_r = \bar{\mathfrak{H}}_r = \mathfrak{G}_r \eta_r$. Thus η_r is an isomorphism of $\mathfrak{G}_r$ onto $\mathfrak{H}_r$. Also λ_r is an isomorphism of $\mathfrak{H}_r = \mathfrak{G}_r \eta_r$ onto $\mathfrak{G}_r$.

Now λ_r maps every element of $\mathfrak{H}_1 \times \cdots \times \mathfrak{H}_{r-1} \times \mathfrak{G}_{r+1} \times \cdots \times \mathfrak{G}_s$ onto 1. Hence, since λ_r induces an isomorphism of $\mathfrak{H}_r$,

$$\mathfrak{H}_r \cap (\mathfrak{H}_1 \cdots \mathfrak{H}_{r-1} \mathfrak{G}_{r+1} \cdots \mathfrak{G}_s) = 1.$$

Hence

$$(39) \qquad \mathfrak{G}' \equiv \mathfrak{H}_1 \cdots \mathfrak{H}_r \mathfrak{G}_{r+1} \cdots \mathfrak{G}_s$$
$$= \mathfrak{H}_1 \times \cdots \times \mathfrak{H}_r \times \mathfrak{G}_{r+1} \times \cdots \times \mathfrak{G}_s.$$

If $x = x_1 x_2 \cdots x_s$, $x_i \, \varepsilon \, \mathfrak{H}_i$ for $i \leq r - 1$, $x_j \, \varepsilon \, \mathfrak{G}_j$ for $j \geq r$, then the mapping

$$\theta: \quad x_1 x_2 \cdots x_s \to x_1 \cdots x_{r-1} (x_r \eta_r) x_{r+1} \cdots x_s$$

is a normal M-endomorphism of $\mathfrak{G}$. Evidently θ is an isomorphism of $\mathfrak{G}$ onto $\mathfrak{G}'$. It follows from Theorem 10 that $\mathfrak{G}' = \mathfrak{G}$. Hence (36) holds also for $k = r$. This completes the proof.

The foregoing inductive argument shows that if the $\mathfrak{H}_i$ are suitably ordered then the normal endomorphism η_i defines an isomorphism of $\mathfrak{G}_i$ onto $\mathfrak{H}_i$. It follows that the mapping μ defined by

$$x\mu = (x_1 x_2 \cdots x_s)\mu = (x_1 \eta_1)(x_2 \eta_2) \cdots (x_s \eta_s),$$

$x_i \, \varepsilon \, \mathfrak{G}_i$, is a normal M-automorphism. Evidently $\mathfrak{G}_i \mu = \mathfrak{H}_i$. Hence we can state the first part of the uniqueness theorem also in the following way:

If (34) and (35) are two decompositions of an M-group with chain conditions into indecomposable factors, then $s = t$ and for a suitable ordering of the $\mathfrak{H}_i$, there exists a normal automorphism μ such that $\mathfrak{G}_i \mu = \mathfrak{H}_i$.

EXERCISES

In the following exercises it is assumed that both chain conditions hold for invariant M-subgroups.

1. Prove that if the center of $\mathfrak{G} = 1$ or if $\mathfrak{G} = \mathfrak{G}^{(1)}$, then $\mathfrak{G}$ has only one decomposition into indecomposable groups.

2. Let $\xi_1, \xi_2, \cdots, \xi_s$ and $\eta_1, \eta_2, \cdots, \eta_s$ be the projections determined by two direct decompositions of $\mathfrak{G}$ into indecomposable groups. Show that, if the order of the η's is suitably chosen, then there exists a normal automorphism μ such that $\eta_i = \mu^{-1} \xi_i \mu$, $i = 1, 2, \cdots, s$.

14. Infinite direct products. We shall consider now some ways of generalizing to an arbitrary number of groups the construction of the direct product of a finite number of groups. In dealing with an arbitrary set of groups we shall find it convenient to suppose that the groups are labelled with subscripts α taken from a certain set J. Also the same group can be counted many times, that is, we do not require that $\mathfrak{G}_\alpha \neq \mathfrak{G}_\beta$ if $\alpha \neq \beta$. Thus we have a set $J = \{\alpha\}$, a collection of subgroups $\{\mathfrak{G}\}$ and a single-valued mapping $\alpha \to \mathfrak{G}_\alpha$ of J onto $\{\mathfrak{G}\}$.

We define first the product set $\widetilde{\prod}_{\alpha \,\varepsilon\, J} \mathfrak{G}_\alpha$ of the $\mathfrak{G}_\alpha$. The elements of this set are the "vectors" $(\cdots g_\alpha \cdots)$ with the property that the element in the "α-place" is in the set $\mathfrak{G}_\alpha$. More precisely, the elements of $\widetilde{\Pi}$ are the single-valued mappings $\alpha \to g_\alpha$ of J that have the property that for each α in J the image element g_α is in the associated group $\mathfrak{G}_\alpha$. Accordingly, if g denotes an element of $\widetilde{\Pi}$, then we can also use the usual functional notation $g(\alpha)$ for the image element g_α.

If J is the set $\{1, 2, 3, \cdots\}$ of positive integers, then $\widetilde{\Pi}$ is the set of sequences $(g_1, g_2, \cdots)$ with the property that $g_i = g(i) \,\varepsilon\, \mathfrak{G}_i$ for all i. We remark also that, if J is arbitrary and all the $\mathfrak{G}_\alpha = \mathfrak{G}$, then $\widetilde{\Pi}$ is the complete set of mappings of J into $\mathfrak{G}$. Following our notation for rings (p. 110) we could also denote this set as $(\mathfrak{G}, J)$.

We now make use of the fact that the $\mathfrak{G}_\alpha$ are groups in introducing component-wise multiplication in $\widetilde{\Pi}$. Thus, if g and $h \,\varepsilon\, \widetilde{\Pi}$, then we define gh by the equation

$$(40) \qquad\qquad (gh)(\alpha) = g(\alpha)h(\alpha).$$

Since $(gh)(\alpha) \,\varepsilon\, \mathfrak{G}_\alpha$, $gh \,\varepsilon\, \widetilde{\Pi}$. It is immediate that $\widetilde{\Pi}$ and this multiplication form a group. The identity element 1 of $\widetilde{\Pi}$ is the function such that $1(\alpha) = 1$ for all α and $g^{-1}(\alpha) = g(\alpha)^{-1}$. If all the $\mathfrak{G}_\alpha$ are M-groups, then we can also regard $\widetilde{\Pi}$ as an M-group. For this purpose we define gm by

$$(41) \qquad\qquad (gm)(\alpha) = g(\alpha)m.$$

It is immediate that this satisfies the basic condition (1). We call the M-group thus obtained the *complete direct product* of the M-groups $\mathfrak{G}_\alpha$.

Now let $\mathfrak{H}$ be any subgroup of the M-group $\tilde{\Pi}$, and consider the mapping of $\mathfrak{H}$ into $\mathfrak{G}_\alpha$ defined by $h \to h(\alpha)$. Evidently by (40) and (41) this mapping is a homomorphism of $\mathfrak{H}$ into $\mathfrak{G}_\alpha$. The image $\mathfrak{H}_\alpha$ is an M-subgroup of $\mathfrak{G}_\alpha$. Now we shall say that $\mathfrak{H}$ is a *subdirect product* of the $\mathfrak{G}_\alpha$ if $\mathfrak{H}_\alpha = \mathfrak{G}_\alpha$ for all α, that is, if the homomorphism $h \to h(\alpha)$ is an onto mapping for every $\alpha \, \varepsilon \, J$. It is clear that $\mathfrak{H}$ is in any case a subdirect product of the image groups $\mathfrak{H}_\alpha$.

Of particular interest is a certain subdirect product that we now define. We consider the totality, which we denote as $\prod_{\alpha \, \varepsilon \, J} \mathfrak{G}_\alpha$, of elements $g \, \varepsilon \, \tilde{\Pi}$ that have the property:

$$g(\alpha) = 1 \quad \text{for all but a finite number of } \alpha.$$

If $g(\alpha) = 1$ for $\alpha \neq \alpha_1, \alpha_2, \cdots, \alpha_m$ and $h(\alpha) = 1$ for $\alpha \neq \beta_1, \beta_2, \cdots, \beta_n$, then $(gh)(\alpha) = 1$ for $\alpha \neq \alpha_1, \cdots, \alpha_m; \beta_1, \cdots, \beta_n$. Hence Π is closed under multiplication. Also it is clear that $1 \, \varepsilon \, \Pi$ and that if $g \, \varepsilon \, \Pi$ then $g^{-1} \, \varepsilon \, \Pi$. Hence Π is a subgroup of $\tilde{\Pi}$.

For any $\gamma \, \varepsilon \, J$ we define $\mathfrak{G}_\gamma'$ to be the subset of elements such that $g(\alpha) = 1$ if $\alpha \neq \gamma$. Then it is evident that $\mathfrak{G}_\gamma'$ is a subgroup of Π and that the mapping $h \to h(\gamma)$ is an isomorphism of $\mathfrak{G}_\gamma'$ onto $\mathfrak{G}_\gamma$. This implies, of course, that for each $\gamma \, \varepsilon \, J$ the mapping $h \to h(\gamma)$ is a homomorphism of Π onto $\mathfrak{G}_\gamma$. Hence Π is a subdirect product of the $\mathfrak{G}_\alpha$. We shall call this particular subdirect product the *direct product* of the $\mathfrak{G}_\alpha$. If J is a finite set (and in this case only), $\Pi = \tilde{\Pi}$.

As in the finite case we can give a characterization of Π in terms of the groups $\mathfrak{G}_\gamma'$. Thus it is easy to see that the $\mathfrak{G}_\gamma'$ are invariant M-subgroups of Π and that

1.
$$\prod_{\alpha \, \varepsilon \, J} \mathfrak{G}_\alpha = [\cup \mathfrak{G}_\alpha'],$$

2.
$$\mathfrak{G}_\beta' \cap [\bigcup_{\alpha \neq \beta} \mathfrak{G}_\alpha'] = 1.$$

Here as usual $[\cup \mathfrak{G}_\alpha']$ denotes the subgroup generated by the groups $\mathfrak{G}_\alpha'$. Conversely, if $\mathfrak{G}$ is any M-group that contains invariant M-subgroups $\mathfrak{G}_\alpha'$ satisfying 1 and 2, then $\mathfrak{G}$ is isomorphic to the direct product of the $\mathfrak{G}_\alpha'$. In this case, too, we shall say simply that $\mathfrak{G}$ is the *direct product* of its subgroups and accordingly we write $\mathfrak{G} = \Pi\mathfrak{G}_\alpha'$.

EXERCISES

1. Let $\mathfrak{G}$ be a commutative group without elements of infinite order. For each prime p let $\mathfrak{G}_p$ be the subset of elements of order a power of p. Show that $\mathfrak{G}_p$ is a subgroup of $\mathfrak{G}$ and that $\mathfrak{G} = \prod_p \mathfrak{G}_p$.

2. Show that, if the group $\mathfrak{G}$ considered in 1 is the additive group of a ring, then the $\mathfrak{G}_p$ are ideals. Hence the ring $\mathfrak{G}$ is the direct sum $\sum_p \oplus \mathfrak{G}_p$* and $\mathfrak{G}_p\mathfrak{G}_q = 0$ if $p \neq q$.

3. Let $\mathfrak{G}$ be an M-group and let $\{\mathfrak{R}_\alpha\}$ be a collection of invariant M-subgroups of $\mathfrak{G}$ such that $\bigcap \mathfrak{R}_\alpha = 1$. Show that $\mathfrak{G}$ is isomorphic to a subdirect product of the groups $\mathfrak{G}_\alpha = \mathfrak{G}/\mathfrak{R}_\alpha$.

* This is the additive terminology and notation that correspond to the direct product Π.

Chapter VI

MODULES AND IDEALS

The concept of a module that we consider in this chapter is a composite notion based on the concepts of a ring and of a group with operators. Modules are of fundamental importance in the study of homomorphisms of abstract rings into rings of endomorphisms of commutative groups (so-called representation theory). This was first recognized by Emmy Noether. Previously the concept of a module had made its appearance in the theory of algebraic numbers.

In the first part of this chapter we introduce the basic module concepts. We investigate further the chain conditions on modules and the related Hilbert basis condition both in the general case and in the special case of ideals. The second part of the chapter is devoted to the derivation of the fundamental decomposition theorems for ideals in Noetherian rings (commutative rings with ascending chain condition). Finally we take up the notion of integral dependence. A special case of this is the concept of algebraic dependence considered in Chapter III. The results that we give here are therefore applicable also in the theory of fields.

1. Definitions

Definition 1. *A* left module *is a commutative group* $\mathfrak{M}$ *(composition addition) with an operator set* $\mathfrak{A}$ *that is a ring such that in addition to the basic operator condition*

$$1_l. \qquad a(x + y) = ax + ay, \quad a \, \varepsilon \, \mathfrak{A}, \quad x,y \, \varepsilon \, \mathfrak{M}$$

we have also

162

$2_l.$
$$(a + b)x = ax + bx$$

and

$3_l.$
$$(ab)x = a(bx).$$

In the present context we employ the notation a_l for the endomorphism $x \to ax$ in the commutative group $\mathfrak{M}$. The conditions 2_l and 3_l are equivalent to the following conditions on these endomorphisms:

$2_l'.$
$$(a + b)_l = a_l + b_l$$

$3_l'.$
$$(ab)_l = b_l a_l.$$

Hence we see that the basic mapping $a \to a_l$ is an anti-homomorphism of $\mathfrak{A}$ into the ring of endomorphisms of $\mathfrak{M}$. Conversely, if $\mathfrak{M}$ is a commutative group with a ring $\mathfrak{A}$ as a set of operators and if the mapping $a \to a_l$ is an anti-homomorphism, then $\mathfrak{M}$ is a left $\mathfrak{A}$-module.

We have seen that the condition 1 implies that

$$(1) \qquad a0 = 0, \quad a(-x) = -ax.$$

Also since $a \to a_l$ is an anti-homomorphism, $0_l = 0$ and $(-a)_l = -a_l$. Hence

$$(2) \qquad 0x = 0, \quad (-a)x = -ax.$$

The concept of a *right module* is defined in a similar fashion. Here we have a commutative group with operator set $\mathfrak{A}$ that is a ring, and we assume that the mapping of $a \, \varepsilon \, \mathfrak{A}$ into the associated endomorphism of $\mathfrak{M}$ is a ring homomorphism. It is convenient to denote the endomorphism associated with a by a_r. Also we denote the product of a in $\mathfrak{A}$ and x in $\mathfrak{M}$ by xa, so that $xa_r = xa$. Then our assumptions on this product can be expressed in the following way:

$1_r.$
$$(x + y)a = xa + ya$$

$2_r.$
$$x(a + b) = xa + xb$$

$3_r.$
$$x(ab) = (xa)b.$$

If $\mathfrak{A}$ is a commutative ring, any homomorphism of $\mathfrak{A}$ is also an anti-homomorphism and conversely. Hence, any left module for

such a ring can be regarded as a right module and conversely.
This is not the case for arbitrary rings. However, if $\mathfrak{A}$ is arbitrary, and $\mathfrak{A}'$ is a ring anti-isomorphic to $\mathfrak{A}$, then any left (right)
$\mathfrak{A}$-module can be regarded as a right (left) $\mathfrak{A}'$-module. For this
purpose we may set $xa' = ax$ ($a'x = xa$) where $a \to a'$ is an anti-isomorphism of $\mathfrak{A}$ onto $\mathfrak{A}'$. Then it is clear that the correspondence
$a' \to a_l$ ($a' \to a_r$) is a homomorphism (anti-homomorphism) of
$\mathfrak{A}'$ as required.

We have seen that the additive group of a ring can be used in a
natural way as the group part of three groups with operators.
In the first of these we take the product of a in the ring $\mathfrak{A}$ by x
in the additive group $\mathfrak{M} = \mathfrak{A},+$ to be the ring product ax.
Evidently 2_l and 3_l hold. Hence this group with operators is a
left module. From now on we shall refer to this module as *the
left module of the ring* $\mathfrak{A}$. Similarly we obtain *the right module
of the ring* $\mathfrak{A}$ by taking $\mathfrak{M} = \mathfrak{A},+$ and defining xa for x in $\mathfrak{M}$
and a in $\mathfrak{A}$ to be the ring product.

2. Fundamental concepts. From now on we deal exclusively
with left modules and we refer to these simply as "modules" or
"$\mathfrak{A}$-modules." It is evident that what we say about these can
also be said about right modules.

Let $\mathfrak{M}$ be an $\mathfrak{A}$-module and let $\mathfrak{N}$ be an $\mathfrak{A}$-subgroup of $\mathfrak{M}$. By
this we mean of course that $\mathfrak{N}$ is a subgroup of $\mathfrak{M}$ and that $\mathfrak{N}$ is
closed under multiplication by elements of $\mathfrak{A}$. Now it is clear
that the product ay, $a \in \mathfrak{A}$, $y \in \mathfrak{N}$ satisfies 2_l and 3_l. Hence $\mathfrak{N}$ is
a module. We call such a module a *submodule* of $\mathfrak{M}$.

If $\mathfrak{N}$ is a submodule of $\mathfrak{M}$, then we know that the factor group
$\mathfrak{M}/\mathfrak{N}$ can be turned into an $\mathfrak{A}$-group by defining

$$a(x + \mathfrak{N}) = ax + \mathfrak{N}.$$

Here again it is immediate that this composition defines a module.
We call this module the *difference module* of $\mathfrak{M}$ relative to $\mathfrak{N}$.
We shall have occasion in the sequel to deal simultaneously with
difference rings and with difference modules. It will therefore
be convenient to adopt the following notational convention:
difference rings will be denoted as before by $\mathfrak{A}/\mathfrak{B}$, difference
modules will be denoted hereafter as $\mathfrak{M} - \mathfrak{N}$.

The concepts of *homomorphism, isomorphism, endomorphism* and *automorphism* for $\mathfrak{A}$-modules are special cases of these concepts for groups with operators. Hence the results that we derived for these notions carry over without alteration to the module case. For example, we know that the image $\mathfrak{M}\eta$ of a module under a homomorphism η is a submodule. Also the kernel $\mathfrak{K}$ of this mapping is a submodule of $\mathfrak{M}$ and we have the "fundamental theorem" that $\mathfrak{M}\eta \cong \mathfrak{M} - \mathfrak{K}$. We know also that the submodules of the left module of the ring $\mathfrak{A}$ are just the left ideals $\mathfrak{J}$.

An important application of these ideas is the definition of the order ideal of an element of a module $\mathfrak{M}$. Let x be any element of $\mathfrak{M}$ and consider the mapping $a \to ax$ of $\mathfrak{A}$ into $\mathfrak{M}$. Evidently this is a group homomorphism. Moreover, since

$$(3) \qquad\qquad ba \to (ba)x = b(ax),$$

it is an $\mathfrak{A}$-homomorphism. We can therefore draw the following conclusions: The set $\mathfrak{A}x$ of image elements ax is a submodule of $\mathfrak{M}$ and the kernel $\mathfrak{J}_x$ of the mapping is a left ideal (submodule) of the ring $\mathfrak{A}$. By definition $\mathfrak{J}_x$ is the set of elements c of $\mathfrak{A}$ such that $cx = 0$. We call this ideal the *order* of the element x. By the fundamental theorem $\mathfrak{A}x \cong \mathfrak{A} - \mathfrak{J}_x$.

We consider next the kernel $\mathfrak{Z}$ of the *ring* anti-homomorphism $a \to a_l$ of $\mathfrak{A}$ into the ring of endomorphisms of $\mathfrak{M}$. The set $\mathfrak{Z}$ is evidently the intersection $\cap\,\mathfrak{J}_x$ of all the order ideals of the elements of $\mathfrak{M}$. The subring $\mathfrak{A}_l$ of image elements a_l is anti-isomorphic to $\mathfrak{A}/\mathfrak{Z}$. We shall call $\mathfrak{Z}$ the *annihilator of the module* $\mathfrak{M}$, and we find it convenient to denote this ideal as $0 : \mathfrak{M}$.

More generally if $\mathfrak{N}_1$ and $\mathfrak{N}_2$ are two submodules of $\mathfrak{M}$, then we denote the set of elements c of $\mathfrak{A}$ such that

$$(4) \qquad\qquad c\mathfrak{N}_2 \subseteq \mathfrak{N}_1$$

by $\mathfrak{N}_1 : \mathfrak{N}_2$. It is immediate that $\mathfrak{N}_1 : \mathfrak{N}_2$ is a (two-sided) ideal in $\mathfrak{A}$. We refer to this ideal as the *quotient of $\mathfrak{N}_1$ by $\mathfrak{N}_2$*. As we shall see later, the study of quotient ideals is of great importance in the ideal theory of commutative rings.

If $\mathfrak{B}$ is a subring of the ring $\mathfrak{A}$, then it is clear that any $\mathfrak{A}$-module can also be regarded as a $\mathfrak{B}$-module. Assume next that $\mathfrak{M}$ is an $\mathfrak{A}$-module and that $\mathfrak{U}$ is an ideal in $\mathfrak{A}$ that is contained in $0 : \mathfrak{M}$.

We shall now show that we can regard $\mathfrak{M}$ also as an $\mathfrak{A}/\mathfrak{U}$-module. Thus let a_1 and a_2 be any two elements of $\mathfrak{A}$ that belong to the same coset mod $\mathfrak{U}$. Then $a_2 = a_1 + u$, u in $\mathfrak{U}$. Hence for any x in $\mathfrak{M}$ we have $a_2 x = a_1 x + ux = a_1 x$. It follows from this that the product defined by

$$(5) \qquad (a + \mathfrak{U})x = ax$$

is single-valued from $\mathfrak{A}/\mathfrak{U} \times \mathfrak{M}$ into $\mathfrak{M}$. It can be verified directly that this composition satisfies 1_l, 2_l and 3_l. Hence we obtain in this way an $\mathfrak{A}/\mathfrak{U}$-module.

EXERCISES

1. If $\mathfrak{J}$ is a left ideal of $\mathfrak{A}$, let $\mathfrak{J}\mathfrak{M}$ denote the set of finite sums $\Sigma b_i x_i$, b_i in $\mathfrak{J}$, x_i in $\mathfrak{M}$. Show that $\mathfrak{J}\mathfrak{M}$ is a submodule of $\mathfrak{M}$.

2. If $\mathfrak{J}$ is a right ideal of $\mathfrak{A}$, the totality of elements $y \, \varepsilon \, \mathfrak{M}$ such that $by = 0$ for all $b \, \varepsilon \, \mathfrak{J}$ is a submodule.

3. Let $\mathfrak{A}$ be a ring with an identity 1. Show that any $\mathfrak{A}$-module permits a decomposition $\mathfrak{M} = 1\mathfrak{M} \oplus \mathfrak{N}$ where $1\mathfrak{M}$ is the submodule of elements $1x$, and $\mathfrak{N}$ is the submodule of elements annihilated by every $a \, \varepsilon \, \mathfrak{A}$.

4. What are the following quotients in the ring of integers: $(6):(3)$, $(6):(15)$, $(3):(9)$?

5. Prove the following rules for quotients: (a) $\mathfrak{N}_1:\mathfrak{N}_2 = \mathfrak{A}$ if $\mathfrak{N}_1 \supseteq \mathfrak{N}_2$, (b) $(\mathfrak{N}_1 \cap \mathfrak{N}_2 \cap \cdots \cap \mathfrak{N}_k):\mathfrak{N} = \mathfrak{N}_1:\mathfrak{N} \cap \mathfrak{N}_2:\mathfrak{N} \cap \cdots \cap \mathfrak{N}_k:\mathfrak{N}$, (c) $\mathfrak{N}_1:\mathfrak{N}_2 = \mathfrak{N}_1:(\mathfrak{N}_1 + \mathfrak{N}_2)$.

6. Show that, if $\mathfrak{N}_1 \subseteq \mathfrak{N}_2$, then $\mathfrak{N}_1:\mathfrak{N}_2 = 0:(\mathfrak{N}_2 - \mathfrak{N}_1)$.

7. Prove that, if $\mathfrak{A}$ is a ring with an identity, then $\mathfrak{J}:\mathfrak{A}$ is the largest two-sided ideal of $\mathfrak{A}$ contained in the left ideal $\mathfrak{J}$.

3. Generators. Unitary modules. If X is a subset of a module $\mathfrak{M}$, then the set (X) of elements of the form

$$(6) \qquad m_1 x_1 + m_2 x_2 + \cdots + m_r x_r + a_1 x_1 + a_2 x_2 + \cdots + a_r x_r$$

where the m_i are integers, the a_i are in $\mathfrak{A}$ and the x_i are in X, is a submodule of $\mathfrak{M}$. Evidently $(X) \supseteq X$ and (X) is contained in every submodule of $\mathfrak{M}$ that contains X. Hence we call (X) the *submodule generated* by X. If $(X) = \mathfrak{M}$, we say that X is a *set of generators* for $\mathfrak{M}$. If there exists a finite set of generators for $\mathfrak{M}$, then we call $\mathfrak{M}$ a *finitely generated* module and, if there exists a single generator, then $\mathfrak{M}$ is a *cyclic* module.

The formula (6) that gives the dependence of an element on a set of generators is somewhat complicated in that it involves coefficients m_i that are integers as well as a_i that are in the ring $\mathfrak{A}$. A simpler formula can be given in the special case of modules that are *unitary* in the sense that $\mathfrak{A}\mathfrak{M} = \mathfrak{M}$. By this equation we mean that every element of $\mathfrak{M}$ can be written in the form $\Sigma a_i y_i$, $a_i \in \mathfrak{A}$, $y_i \in \mathfrak{M}$. We shall now prove the following

Theorem 1. *If X is a set of generators for a unitary module $\mathfrak{M}$, then every element of $\mathfrak{M}$ can be written in the form*

$$(7) \qquad a_1 x_1 + a_2 x_2 + \cdots + a_r x_r$$

where the $a_i \in \mathfrak{A}$ and the $x_i \in X$.

Proof. Let x be any element of $\mathfrak{M}$ and write $x = \Sigma a_i y_i$ for suitable a_i in $\mathfrak{A}$, y_i in $\mathfrak{M}$. Then there exist elements x_j in X such that

$$y_i = \Sigma m_{ij} x_j + \Sigma a_{ij} x_j, \quad m_{ij} \in I, \quad a_{ij} \in \mathfrak{A}.$$

Then $x = \Sigma a_i y_i = \Sigma m_{ij} a_i x_j + \Sigma a_i a_{ij} x_j = \Sigma b_j x_j$ where

$$b_j = \sum_i m_{ij} a_i + \sum_i a_i a_{ij}.$$

In particular, we see that, if $\mathfrak{M}$ is cyclic and unitary, then $\mathfrak{M}$ contains an element x such that every element of $\mathfrak{M}$ is a multiple ax of x. In particular, x has the form ex for a suitable e in $\mathfrak{A}$. If $\mathfrak{M}$ is unitary and $\mathfrak{A}$ has an identity 1, then 1 acts as identity operator for $\mathfrak{M}$; for, if $x = \Sigma a_i y_i$, then $1x = 1(\Sigma a_i y_i) = \Sigma(1a_i)y_i = \Sigma a_i y_i = x$. Conversely, it is clear that, if 1 acts as identity operator, then any x has the form $1x$ so that $\mathfrak{M}$ is unitary. Thus, *if $\mathfrak{A}$ has an identity, then the condition that $\mathfrak{M}$ be unitary is equivalent to the condition that 1_l is the identity mapping in $\mathfrak{M}$.*

A unitary module for which the basic ring $\mathfrak{A}$ is a division ring is called a *vector space*. The detailed study of vector spaces constitutes the subject matter of Volume II of these *Lectures*.

EXERCISES

1. Call a left ideal $\mathfrak{J}$ *regular* if there exists an element e such that $xe \equiv x$ mod $\mathfrak{J}$ holds for all x in $\mathfrak{A}$. Prove that, if $\mathfrak{M}$ is a unitary cyclic module, then $\mathfrak{M} \cong \mathfrak{A} - \mathfrak{J}$ where $\mathfrak{J}$ is a suitable regular left ideal.

2. Prove that, if $\mathfrak{J}$ is regular, then $\mathfrak{J} \supseteq \mathfrak{J}{:}\mathfrak{A}$.

3. Let $\mathfrak{M}$ be a simple $\mathfrak{A}$-module. Prove that either $\mathfrak{A}\mathfrak{M} = 0$ in which case $\mathfrak{M}$ is finite and has a prime number of elements, or $\mathfrak{M}$ is a unitary cyclic module

with any non-zero element as generator. Show that conversely either of these conditions insures that $\mathfrak{M}$ is simple. (Note that the first part of this exercise is a generalization of ex. 1, p. 78.)

4. The chain conditions.

The chain conditions that were introduced for groups with operators play an important role in various aspects of module and ideal theory. As we shall see (next section) the ideals in a polynomial ring over a field satisfy the ascending chain condition and this fact alone suffices for the derivation of the basic ideal decomposition theorems for such a ring. On the other hand, the study of rings that satisfy the descending chain condition for ideals forms an important part of the so-called structure theory of rings.

In this section and the next we shall derive some of the simpler implications of the chain conditions. We note first that, since any module is a commutative group, the chain conditions for modules can be stated in the following way:

Descending chain condition. If $\mathfrak{N}_1 \supseteq \mathfrak{N}_2 \supseteq \cdots$ is a decreasing sequence of submodules, then there exists an integer N such that $\mathfrak{N}_N = \mathfrak{N}_{N+1} = \cdots$.

Ascending chain condition. If $\mathfrak{N}_1 \subseteq \mathfrak{N}_2 \subseteq \cdots$ is an increasing sequence of submodules, then there exists an integer N such that $\mathfrak{N}_N = \mathfrak{N}_{N+1} = \cdots$.

It is easy to see (using the axiom of choice) that the descending chain condition is equivalent to the

Minimum condition. In any non-vacuous collection $\{\mathfrak{N}\}$ of submodules, there exists a minimal submodule, that is, a submodule that does not contain properly any submodule of the collection.

To establish this equivalence we assume first the descending chain condition. Let $\{\mathfrak{N}\}$ be a non-vacuous collection of submodules. Select $\mathfrak{N}_1$ in the collection. Either $\mathfrak{N}_1$ is minimal or there is an $\mathfrak{N}_2$ in $\{\mathfrak{N}\}$ such that $\mathfrak{N}_2 \subset \mathfrak{N}_1$. Either $\mathfrak{N}_2$ is minimal or there is an $\mathfrak{N}_3$ in $\{\mathfrak{N}\}$ such that $\mathfrak{N}_3 \subset \mathfrak{N}_2$. This process leads in a finite number of steps to a minimal submodule; for otherwise, by the axiom of choice, we obtain an infinite chain $\mathfrak{N}_1 \supset \mathfrak{N}_2 \supset \mathfrak{N}_3 \supset \cdots$ contrary to assumption. Conversely, suppose that the minimum condition holds, and let $\mathfrak{N}_1 \supseteq \mathfrak{N}_2 \supseteq \cdots$ be an infinite decreasing sequence of submodules. Let $\mathfrak{N}_N$ be a minimal element in the collection $\{\mathfrak{N}_i\}$. Then we certainly have $\mathfrak{N}_N = \mathfrak{N}_{N+1} = \cdots$.

In a similar manner we can show that the ascending chain condition is equivalent to the

Maximum condition. Any non-vacuous collection of submodules contains a maximum submodule (one not contained properly in any other module of the collection).

The maximum condition implies the following useful principle of induction: Let P be a property of submodules of a module such that $P(\mathfrak{N})$ holds if $P(\mathfrak{N}')$ holds for every $\mathfrak{N}' \supset \mathfrak{N}$. Then $P(\mathfrak{N})$ is true for all $\mathfrak{N}$. As in the case of the second principle of induction for natural numbers (p. 9), the proof follows directly from the consideration of the collection of submodules $\mathfrak{N}$ such that $P(\mathfrak{N})$ is false.

The next result that we shall derive is very useful in the theory of ideals. We state it as the following

Theorem 2. *A module $\mathfrak{M}$ satisfies the ascending chain condition for submodules if and only if every submodule of $\mathfrak{M}$ is finitely generated.*

Proof. We assume first that the ascending chain condition holds and we let $\mathfrak{N}$ be any submodule of $\mathfrak{M}$. If $\mathfrak{N} = 0$, then $\mathfrak{N}$ is generated by 0. If $\mathfrak{N} \neq 0$, let u_1 be any non-zero element of $\mathfrak{N}$ and let (u_1) denote the submodule generated by u_1. If $(u_1) \subset \mathfrak{N}$, let $u_2 \,\varepsilon\, \mathfrak{N}$, $\notin (u_1)$. Then the submodule (u_1, u_2) generated by u_1, u_2 properly contains (u_1). If $(u_1, u_2) \subset \mathfrak{N}$, we can find a u_3 in $\mathfrak{N}$ such that $(u_1, u_2, u_3) \supset (u_1, u_2)$. After a finite number of selections we obtain $(u_1, u_2, \cdots, u_n) = \mathfrak{N}$, since otherwise we obtain an infinite properly ascending chain of submodules $(u_1) \subset (u_1, u_2) \subset (u_1, u_2, u_3) \subset \cdots$.

We assume next that any submodule is finitely generated and we let $\mathfrak{N}_1 \subseteq \mathfrak{N}_2 \subseteq \mathfrak{N}_3 \subseteq \cdots$ be an arbitrary ascending chain of submodules. The proof that $\mathfrak{N}_N = \mathfrak{N}_{N+1} = \cdots$ for some N is similar to the proof of the ascending chain condition for principal ideal domains (p. 121). As in the special case we note first that the logical sum $\mathfrak{P} = \cup \mathfrak{N}_i$ is a submodule. Hence $\mathfrak{P} = (u_1, u_2, \cdots, u_r)$ for suitable u_i in $\mathfrak{P}$. Now $u_i \,\varepsilon\, \mathfrak{N}_{h_i}$ for some h_i. If $N =$

max $(h_1, h_2, \cdots, h_r)$, then every $u_i \,\varepsilon\, \mathfrak{N}_N$. Hence $\mathfrak{P} \subseteq \mathfrak{N}_N$ and this evidently implies that $\mathfrak{N}_N = \mathfrak{N}_{N+1} = \cdots$.

5. The Hilbert basis theorem. We suppose now that $\mathfrak{M}$ is a finitely generated unitary module. We shall prove that, if the ring $\mathfrak{A}$ satisfies the ascending (descending) chain condition for left ideals, then the same condition holds for $\mathfrak{M}$.

Let $x_1, x_2, \cdots, x_r$ be a fixed set of generators for $\mathfrak{M}$. Then if $\mathfrak{N}$ is any submodule of $\mathfrak{M}$, we define the subset $\mathfrak{I}_j\,(\mathfrak{N})$ of $\mathfrak{A}, j = 1, 2, \cdots, r$, to be the totality of elements b for which there exists an element

$$b x_j + b_{j+1} x_{j+1} + \cdots + b_r x_r$$

in $\mathfrak{N}$. It is immediate that $\mathfrak{I}_j(\mathfrak{N})$ is a left ideal. Moreover, we evidently have $\mathfrak{I}_j(\mathfrak{N}) \subseteq \mathfrak{I}_j(\mathfrak{P})$ for all j if $\mathfrak{N}$ is contained in the submodule $\mathfrak{P}$. We note next the following

Lemma 1. If $\mathfrak{N} \subseteq \mathfrak{P}$ and $\mathfrak{I}_j(\mathfrak{N}) = \mathfrak{I}_j(\mathfrak{P})$ for all j, then $\mathfrak{N} = \mathfrak{P}$.

Proof. Let $y = b_1 x_1 + b_2 x_2 + \cdots + b_r x_r$ be any element of $\mathfrak{P}$. Then $b_1 \,\varepsilon\, \mathfrak{I}_1(\mathfrak{P}) = \mathfrak{I}_1(\mathfrak{N})$. Hence, there is an element y' in $\mathfrak{N}$ of the form $b_1 x_1 + b_2' x_2 + \cdots + b_r' x_r$. Then $y - y' = c_2 x_2 + c_3 x_3 + \cdots + c_r x_r$ where $c_i = b_i - b_i'$ and $y - y' \,\varepsilon\, \mathfrak{P}$. Hence $c_2 \,\varepsilon\, \mathfrak{I}_2(\mathfrak{P}) = \mathfrak{I}_2(\mathfrak{N})$. Now there is an element y'' in $\mathfrak{N}$ of the form $c_2 x_2 + c_3' x_3 + \cdots + c_r' x_r$. Then $y - y' - y'' = d_3 x_3 + \cdots + d_r x_r$. Continuing in this way, we obtain $y', y'', \cdots, y^{(r)}$ in $\mathfrak{N}$ such that $y - y' - \cdots - y^{(r)} = 0$. Hence $y = y' + y'' + \cdots + y^{(r)} \,\varepsilon\, \mathfrak{N}$.

Now let $\mathfrak{N}_1 \subseteq \mathfrak{N}_2 \subseteq \cdots$ be an increasing chain of submodules of $\mathfrak{M}$. Then we can associate with this chain the r chains of left ideals

$$\mathfrak{I}_j(\mathfrak{N}_1) \subseteq \mathfrak{I}_j(\mathfrak{N}_2) \subseteq \cdots, \quad j = 1, 2, \cdots, r.$$

If the ascending chain condition holds in $\mathfrak{A}$, we can find for each j an integer N_j such that

$$\mathfrak{I}_j(\mathfrak{N}_{N_j}) = \mathfrak{I}_j(\mathfrak{N}_{N_j+1}) = \cdots, \quad j = 1, 2, \cdots, r.$$

Hence, if $N = \max (N_1, N_2, \cdots, N_r)$, then $\mathfrak{I}_j(\mathfrak{N}_N) = \mathfrak{I}_j(\mathfrak{N}_{N+1}) = \cdots$ holds for all j. By Lemma 1 this implies that $\mathfrak{N}_N = \mathfrak{N}_{N+1} = \cdots$. We have therefore proved the "ascending chain" part of the following

Theorem 3. *If $\mathfrak{A}$ is a ring that satisfies the ascending (descending) chain condition for left ideals, then any finitely generated unitary $\mathfrak{A}$-module $\mathfrak{M}$ satisfies the ascending (descending) chain condition for submodules.*

The proof of this result for descending chains is similar to the above.

We wish to show next that, if $\mathfrak{A}$ is a ring with an identity that satisfies the ascending chain condition or, equivalently, if every left ideal in $\mathfrak{A}$ is finitely generated, then the same condition holds for the polynomial ring $\mathfrak{A}[x]$ in a transcendental element x. The proof of this result is quite similar to the foregoing.

With each left ideal $\mathfrak{N}$ of $\mathfrak{A}[x]$ and each $j = 0, 1, 2, \cdots$ we associate the set $\mathfrak{J}_j(\mathfrak{N})$ of elements $b \, \varepsilon \, \mathfrak{A}$ such that there exists an element

$$bx^j + b_{j-1}x^{j-1} + \cdots + b_0$$

in $\mathfrak{N}$. Then it is clear that $\mathfrak{J}_j(\mathfrak{N})$ is a left ideal in $\mathfrak{A}$. Also if $bx^j + b_{j-1}x^{j-1} + \cdots + b_0 \, \varepsilon \, \mathfrak{N}$, then so does

$$bx^{j+1} + b_{j-1}x^j + \cdots + b_0x = x(bx^j + b_{j-1}x^{j-1} + \cdots + b_0).$$

Hence

$$\mathfrak{J}_0(\mathfrak{N}) \subseteq \mathfrak{J}_1(\mathfrak{N}) \subseteq \mathfrak{J}_2(\mathfrak{N}) \subseteq \cdots .$$

Consequently the set $\mathfrak{J}(\mathfrak{N}) = \cup \mathfrak{J}_j(\mathfrak{N})$ is a left ideal. We shall now use these remarks in proving the important

Hilbert basis theorem. *Let $\mathfrak{A}$ be a ring with an identity that has the property that every left ideal in $\mathfrak{A}$ is finitely generated. Then the ring $\mathfrak{A}[x]$ of polynomials in a transcendental element x also has this property.*

Proof. Let $\mathfrak{N}$ be an ideal and define the ideals $\mathfrak{J}_j(\mathfrak{N})$ and $\mathfrak{J}(\mathfrak{N})$ as above. Then there is an integer N such that $\mathfrak{J}_N(\mathfrak{N}) = \mathfrak{J}_{N+1}(\mathfrak{N}) = \cdots = \mathfrak{J}(\mathfrak{N})$. Let b_{ji}, $j = 0, 1, 2, \cdots, N$; $i = 1, 2, \cdots, m_j$ be elements of $\mathfrak{A}$ such that

$$\mathfrak{J}_j(\mathfrak{N}) = (b_{j1}, b_{j2}, \cdots, b_{jm_j})$$

and let $f_{ji}(x)$ be polynomials in $\mathfrak{N}$ such that

$$f_{ji}(x) = b_{ji}x^j + c_{ji}x^{j-1} + d_{ji}x^{j-2} + \cdots .$$

Then we shall show that $\mathfrak{N} = (f_{01}, \cdots, f_{0m_0}; f_{11}, \cdots, f_{1m_1}; \cdots; \cdots f_{N,m_N})$. Thus let $g = c_r x^r + c_{r-1} x^{r-1} + \cdots \varepsilon \mathfrak{N}$. If $r \le N$, $c_r = a_{r1} b_{r1} + a_{r2} b_{r2} + \cdots + a_{rm_r} b_{rm_r}$ for suitable a_{ri} in $\mathfrak{A}$. Hence $g - \Sigma a_{ri} f_{ri}(x)$ is a polynomial in $\mathfrak{N}$ of degree $< r$. If $r > N$, $c_r = a_{r1} b_{N1} + a_{r2} b_{N2} + \cdots + a_{rm_N} b_{Nm_N}$, a_{ri} in $\mathfrak{A}$; hence $g - \Sigma a_{ri} x^{r-N} f_{Ni}(x)$ is a polynomial in $\mathfrak{N}$ of degree $< r$. We can therefore reach our conclusion by using induction on the degree of g.

Hilbert's theorem has an immediate extension to polynomials in several elements. The result is the following

Corollary 1. *Let $\mathfrak{A}$ be a ring with an identity such that every left ideal in $\mathfrak{A}$ is finitely generated. Then every left ideal in $\mathfrak{A}[x_1, x_2, \cdots, x_r]$ has a finite set of generators.*

An important special case of this result is the

Corollary 2. *If $\mathfrak{A}$ is a division ring or if $\mathfrak{A}$ is a principal ideal domain, then every left (right) ideal of $\mathfrak{A}[x_1, x_2, \cdots, x_r]$ has a finite set of generators.*

EXERCISES

1. Prove that, insofar as the ascending chain condition is concerned, the assumption that $\mathfrak{M}$ is unitary is superfluous in Theorem 3.

2. Prove that, if $\mathfrak{A}$ has an identity and every left ideal of $\mathfrak{A}$ is finitely generated, then every left ideal in the ring $\mathfrak{A}\langle x \rangle$ of power series in x (defined in ex. 1, p. 95) is finitely generated.

3. Let $\mathfrak{F}$ be a finite field of q elements and let $\mathfrak{B}$ be the ideal in $\mathfrak{F}[x_1, x_2, \cdots, x_r]$ of polynomials $m(x_1, \cdots, x_r)$ such that $m(s_1, \cdots, s_r) = 0$ for all s_i in $\mathfrak{F}$. Determine a finite set of generators for $\mathfrak{B}$.

6. Noetherian rings. Prime and primary ideals. In the next few sections we shall develop the basic results of the theory of ideals in commutative rings with ascending chain condition. We have seen that this class of rings includes the polynomial rings $\mathfrak{F}[x_1, x_2, \cdots, x_r]$ where $\mathfrak{F}$ is a field. The theory of polynomial ideals is fundamental in algebraic geometry. The abstract development of this theory on the basis only of the ascending chain condition and commutativity was initiated by Emmy Noether. For this reason one calls a ring that satisfies these two conditions a *Noetherian ring*.

We assume first only that $\mathfrak{A}$ is commutative. In the case of principal ideal domains we have seen that an element d is a divisor of an element b if and only if the ideal $(d) \supseteq (b)$. For this reason if $\mathfrak{D}$ and $\mathfrak{B}$ are ideals in any commutative ring, then we say that $\mathfrak{D}$ is a *divisor* of $\mathfrak{B}$ and $\mathfrak{B}$ is a *multiple* of $\mathfrak{D}$ if $\mathfrak{D} \supseteq \mathfrak{B}$. Similarly we are motivated by the principal ideal case in calling $\mathfrak{B}_1 + \mathfrak{B}_2$ the *greatest common divisor* of $\mathfrak{B}_1$ and $\mathfrak{B}_2$ and $\mathfrak{B}_1 \cap \mathfrak{B}_2$ the *least common multiple* of $\mathfrak{B}_1$ and $\mathfrak{B}_2$; for in a principal ideal domain $(b_1) + (b_2) = (d)$ where d is a g.c.d. for b_1 and b_2 and $(b_1) \cap (b_2) = (m)$ where m is a l.c.m. for b_1 and b_2. We generalize next the notion of a prime in the following important

Definition 2. *An ideal* $\mathfrak{B}$ *of a commutative ring* $\mathfrak{A}$ *is* prime *if* $ab \equiv 0 \ (mod \ \mathfrak{B})$ *implies that either* $a \equiv 0 \ (mod \ \mathfrak{B})$ *or* $b \equiv 0 \ (mod \ \mathfrak{B})$.

It is clear that this is equivalent to the condition that $\mathfrak{A}/\mathfrak{B}$ is an integral domain. Also evidently $\mathfrak{A}$ is an integral domain if and only if 0 is a prime ideal. The element p is a prime in the sense of the definition given in Chapter IV if and only if (p) is a prime ideal. Thus, for example, $(x - y)$ is a prime ideal in $\mathfrak{F}[x,y]$. An example of a prime ideal that is not principal is the ideal $(x,y) = (x) + (y)$ in $\mathfrak{F}[x,y]$. Here $\mathfrak{F}[x,y]/(x,y) \cong \mathfrak{F}$.

Any maximal ideal $\mathfrak{B}$ in a ring with an identity is a prime; for, in this case, $\mathfrak{A}/\mathfrak{B}$ is a field and hence also an integral domain. If $\mathfrak{A}$ does not have an identity and $\mathfrak{B}$ is maximal, either $\mathfrak{A}/\mathfrak{B}$ is a field or $(\mathfrak{A}/\mathfrak{B})^2 = 0$. In the first case $\mathfrak{B}$ is prime while in the second we have $\mathfrak{A}^2 \subseteq \mathfrak{B}$.

Suppose next that $\mathfrak{B}$ is any ideal in the commutative ring $\mathfrak{A}$, and let $\mathfrak{R} = \mathfrak{R}(\mathfrak{B})$ be the totality of elements z for which there exists a positive integer r (possibly depending on z) such that $z^r \equiv 0 \ (\text{mod } \mathfrak{B})$. Evidently $\mathfrak{R}$ can also be defined as the set of elements z such that the coset $\bar{z} = z + \mathfrak{B}$ is nilpotent in $\mathfrak{A}/\mathfrak{B}$. We now show that $\mathfrak{R}$ is an ideal. First if $z^r \equiv 0 \ (\text{mod } \mathfrak{B})$ and a is any element of $\mathfrak{A}$, then $(az)^r = a^r z^r \equiv 0 \ (\text{mod } \mathfrak{B})$. Next let z_1 and $z_2 \ \varepsilon \ \mathfrak{R}$ and let $z_i^{r_i} \equiv 0 \ (\text{mod } \mathfrak{B}), i = 1, 2$. Set $r = r_1 + r_2 - 1$. Then

$$(z_1 - z_2)^r = \Sigma m_{ij} z_1{}^i z_2{}^j, \quad i + j = r, \quad m_{ij} \ \varepsilon \ I.$$

In each term we have either $i \geq r_1$ or $j \geq r_2$. Hence $m_{ij} z_1{}^i z_2{}^j \equiv 0 \ (\text{mod } \mathfrak{B})$. Thus $(z_1 - z_2)^r \equiv 0 \ (\text{mod } \mathfrak{B})$ and $z_1 - z_2 \ \varepsilon \ \mathfrak{R}$. This proves our assertion. The ideal $\mathfrak{R} = \mathfrak{R}(\mathfrak{B})$ is called the *(nil)* *radical* of $\mathfrak{B}$. Evidently $\mathfrak{R}$ is a divisor of $\mathfrak{B}$.

Examples. (1) Let $a = p_1^{e_1}p_2^{e_2} \cdots p_r^{e_r}$ be a factorization of the integer a into a product of prime powers $p_i^{e_i}$ where $p_i \neq p_j$ if $i \neq j$. Then the radical of (a) is $(p_1p_2 \cdots p_r)$; for if $b = kp_1p_2 \cdots p_r$ and $e = \max(e_1, e_2, \cdots, e_r)$, then $b^e \equiv 0 \pmod{(a)}$. On the other hand, if a power of c is divisible by a, then c itself is divisible by $p_1 p_2 \cdots p_r$. (2) Consider the ideal (x^2, y^3) in $\mathfrak{F}[x,y]$. Evidently the radical contains x and y. On the other hand, if $f(x,y)^r \equiv 0 \pmod{(x^2, y^3)}$, then the constant term of $f(x,y)$ is 0. Hence $f(x,y) \equiv 0 \pmod{(x,y)}$. Thus the radical of (x^2, y^3) is (x,y).

The radical of an ideal in a Noetherian ring is *nilpotent* modulo this ideal. By this we mean that there exists an integer N such that $\mathfrak{R}^N \equiv 0 \pmod{\mathfrak{B}}$. In order to prove this we choose a finite set of generators $z_1, z_2, \cdots, z_m$ for $\mathfrak{R}$, so that $\mathfrak{R} = (z_1, z_2, \cdots, z_m)$. Let r_i be an integer such that $z_i^{r_i} \equiv 0 \pmod{\mathfrak{B}}$ and set $N = r_1 + r_2 + \cdots + r_m - (m - 1)$. Consider the product of any N elements of $\mathfrak{R}$. Since any element of $\mathfrak{R}$ has the form $\Sigma a_i z_i + \Sigma m_i z_i$, $a_i \in \mathfrak{A}$, $m_i \in I$, such a product has the form

$$\Sigma A_{i_1 \cdots i_m} z_1^{i_1} z_2^{i_2} \cdots z_m^{i_m} + \Sigma M_{i_1 \cdots i_m} z_1^{i_1} \cdots z_m^{i_m}$$

where the A's are in $\mathfrak{A}$, the M's are integers and $i_1 + i_2 + \cdots + i_m = N$. Now it is easy to see that for each term we must have $i_j \geq r_j$ for some j. It follows that this term is in $\mathfrak{B}$. Hence any product of N elements of $\mathfrak{R}$ is in $\mathfrak{B}$ and this implies that $\mathfrak{R}^N \equiv 0$, $\pmod{\mathfrak{B}}$.

We consider next the generalization of the notion of prime-power element in a principal ideal domain. There are several possibilities for such a generalization, but the "right" one for the purposes of the decomposition theory is the one given in the following important

Definition 3. *An ideal $\mathfrak{B}$ in a commutative ring is a primary ideal if every zero divisor modulo $\mathfrak{B}$ is in the radical, that is, if $ab \equiv 0 \pmod{\mathfrak{B}}$ and $b \not\equiv 0 \pmod{\mathfrak{B}}$ implies that $a \equiv 0 \pmod{\mathfrak{R}}$.*

It is a simple consequence of this definition that the radical of a primary ideal is a prime ideal. For let ab be in the radical $\mathfrak{R}$ and suppose $a \not\equiv 0 \pmod{\mathfrak{R}}$. Then $a^r b^r = (ab)^r \equiv 0 \pmod{\mathfrak{B}}$ for some positive integer r. On the other hand, $a^r \not\equiv 0 \pmod{\mathfrak{B}}$. Hence by definition $b^r \equiv 0 \pmod{\mathfrak{R}}$ and this means that $b^{rs} = (b^r)^s \equiv 0 \pmod{\mathfrak{B}}$ for some s. Hence, b is in $\mathfrak{R}$. The radical of a primary ideal is called its *associated prime ideal.*

It is easy to see that (q) is primary in the ring of integers if and only if $q = p^e$, p a prime (ex. 1 below). We leave it to the reader to verify also that the ideal (x^2, y^3) is primary in $\mathfrak{F}[x,y]$. On the other hand, we note that the ideal (x^2, xy) is not primary in $\mathfrak{F}[x,y]$ even though its radical (x) is prime. For $x \not\equiv 0$ (mod (x^2, xy)) and $y \not\equiv 0$ (mod (x)) but $xy \equiv 0$ (mod (x^2, xy)).

EXERCISES

1. Show that (q), $q \neq 0, 1$, is primary in I if and only if $q = p^e$, p a prime.

2. Prove that, if $\mathfrak{B}$ is a prime ideal and $\mathfrak{C}_1$ and $\mathfrak{C}_2$ are ideals such that $\mathfrak{C}_1\mathfrak{C}_2 \equiv 0$ (mod $\mathfrak{B}$), then either $\mathfrak{C}_1 \equiv 0$ (mod $\mathfrak{B}$) or $\mathfrak{C}_2 \equiv 0$ (mod $\mathfrak{B}$).

3. Prove that $\mathfrak{R}(\mathfrak{B}_1 \cap \mathfrak{B}_2) = \mathfrak{R}(\mathfrak{B}_1) \cap \mathfrak{R}(\mathfrak{B}_2)$.

4. Prove that in a Noetherian ring $\mathfrak{B}_1{}^r \subseteq \mathfrak{B}_2$ holds if and only if $\mathfrak{R}(\mathfrak{B}_1) \subseteq \mathfrak{R}(\mathfrak{B}_2)$.

7. Representation of an ideal as intersection of primary ideals. The fundamental factorization theorem in the ring of integers can be stated in terms of ideals as follows: Every ideal (a) can be written in one and only one way as a product of prime ideals. This does not hold for arbitrary Noetherian rings. A somewhat weaker statement is that every ideal in I is an intersection (least common multiple) of primary ideals; for if $a = p_1{}^{e_1} p_2{}^{e_2} \cdots p_r{}^{e_r}$ where the p_i are distinct primes, then clearly

$$(a) = (p_1{}^{e_1}) \cap (p_2{}^{e_2}) \cap \cdots \cap (p_r{}^{e_r}).$$

We shall show in this section that this type of decomposition is valid in any Noetherian ring. The question of uniqueness will be taken up in § 8.

Assume now that $\mathfrak{A}$ is any Noetherian ring. We shall show first that an ideal that is not primary is *reducible* in the sense that it can be expressed as an intersection of proper divisors. Thus suppose that $\mathfrak{B}$ is not primary and let d be an element which is a zero-divisor modulo $\mathfrak{B}$ but which does not belong to $\mathfrak{R}(\mathfrak{B})$. Let a be an element such that $ad \equiv 0$ (mod $\mathfrak{B}$) and $a \not\equiv 0$ (mod $\mathfrak{B}$). Then $a \in \mathfrak{B}:(d)$, $\notin \mathfrak{B}$. Hence $\mathfrak{B}:(d) \supset \mathfrak{B}$. Also since $d \notin \mathfrak{R}(\mathfrak{B})$, $(d^k) + \mathfrak{B} \supset \mathfrak{B}$ for $k = 1, 2, 3, \cdots$. Consider now the ascending chain

$$(8) \qquad \mathfrak{B}:(d) \subseteq \mathfrak{B}:(d^2) \subseteq \mathfrak{B}:(d^3) \subseteq \cdots.$$

Let r be a positive integer such that

$$(9) \qquad \mathfrak{B}:(d^r) = \mathfrak{B}:(d^{r+1}) = \cdots.$$

Then we have the relation

$$(10) \qquad \mathfrak{B} = (\mathfrak{B}:(d^r)) \cap (\mathfrak{B} + (d^{r+1}));$$

for if $u \in \mathfrak{B} + (d^{r+1})$, $u = b + md^{r+1} + cd^{r+1}$ where $b \in \mathfrak{B}$, $m \in I$, $c \in \mathfrak{A}$. Hence, if $u \in \mathfrak{B}:(d^r)$, then

$$ud^r = bd^r + md^{2r+1} + cd^{2r+1} \equiv 0 \; (\mathrm{mod}\; \mathfrak{B}).$$

This gives $(md + cd)d^{2r} \equiv 0 \; (\mathrm{mod}\; \mathfrak{B})$ so that $md + cd$ is in $\mathfrak{B}:(d^{2r})$. But then, by (9), $(md + cd)d^r \equiv 0 \; (\mathrm{mod}\; \mathfrak{B})$. Hence $md^{r+1} + cd^{r+1} \equiv 0 \; (\mathrm{mod}\; \mathfrak{B})$. Thus $u \in \mathfrak{B}$. This proves (10). Since both ideals in (10) properly contain $\mathfrak{B}$, $\mathfrak{B}$ is reducible. Evidently the result that we have proved can also be stated in the following form:

Theorem 4. *Every irreducible ideal in a Noetherian ring is primary.*

We shall prove next that every ideal in a Noetherian ring is a finite intersection of irreducible ideals. To prove this we use the principle of induction formulated in § 4, that is, we show that for a given ideal $\mathfrak{B}$ the result holds, provided that it holds for all $\mathfrak{B}_1 \supset \mathfrak{B}$. Now either $\mathfrak{B}$ is *irreducible*, in which case we are through, or $\mathfrak{B} = \mathfrak{B}_1 \cap \mathfrak{B}_2$ where $\mathfrak{B}_i \supset \mathfrak{B}$ for $i = 1, 2$. Then $\mathfrak{B}_1$ and $\mathfrak{B}_2$ can be represented as intersections of finite numbers of irreducible ideals. Hence $\mathfrak{B}$, too, is such an intersection. In view of Theorem 4 this result implies the fundamental decomposition theorem:

Theorem 5. *Every ideal in a Noetherian ring is a finite intersection of primary ideals.*

EXERCISES

1. Express (x^2, xy) as a finite intersection of primary ideals.
2. Show that the ideal (x^2, xy, y^2) is primary and reducible in $\mathfrak{F}[x, y]$.
3. (Fitting.) Let $\mathfrak{M}$ be an $\mathfrak{A}$-module ($\mathfrak{A}$ arbitrary) that satisfies the ascending chain condition. Suppose that there exists an $\mathfrak{A}$-endomorphism θ of $\mathfrak{M}$ that

is not nilpotent and that is not an isomorphism of $\mathfrak{M}$. Prove that there exists two submodules $\mathfrak{M}_i \neq 0$ in $\mathfrak{M}$ such that $\mathfrak{M}_1 \cap \mathfrak{M}_2 = 0$.

4. (Fitting.) Let $\mathfrak{M}$ be an $\mathfrak{A}$-module satisfying the ascending chain condition. Suppose that the intersection of any two non-zero modules of $\mathfrak{M}$ is $\neq 0$. Prove that the set of nilpotent $\mathfrak{A}$-endomorphisms of $\mathfrak{M}$ is an ideal $\mathfrak{R}$ in the ring $\mathfrak{E}$ of $\mathfrak{A}$-endomorphisms. Prove that if α in $\mathfrak{E}$ is a left-zero divisor, then $\alpha \, \varepsilon \, \mathfrak{R}$.

8. Uniqueness theorems. We shall say that the ideal $\mathfrak{B}$ is an *irredundant intersection* of ideals $\mathfrak{Q}_1, \mathfrak{Q}_2, \cdots, \mathfrak{Q}_r$ if $\mathfrak{B} = \mathfrak{Q}_1 \cap \mathfrak{Q}_2 \cap \cdots \cap \mathfrak{Q}_r$ and

$$\mathfrak{Q}_1 \cap \cdots \cap \mathfrak{Q}_{i-1} \cap \mathfrak{Q}_{i+1} \cap \cdots \cap \mathfrak{Q}_r \supset \mathfrak{B}$$

for $i = 1, 2, \cdots, r$. It is evident that, if we have any representation of $\mathfrak{B}$ as a finite intersection of ideals, we can omit enough terms to obtain an irredundant intersection. In particular, we see that every ideal in a Noetherian ring is an irredundant intersection of primary ideals. We observe next that it is sometimes possible to combine primary ideals to obtain primary ideals, for we have the following

Lemma 1. *If $\mathfrak{Q}_1$ and $\mathfrak{Q}_2$ are primary ideals that have the same radical $\mathfrak{P}$, then $\mathfrak{Q}_1 \cap \mathfrak{Q}_2$ is primary.*

Proof. We know that $\mathfrak{R}(\mathfrak{Q}_1 \cap \mathfrak{Q}_2) = \mathfrak{R}(\mathfrak{Q}_1) \cap \mathfrak{R}(\mathfrak{Q}_2)$. Hence $\mathfrak{R}(\mathfrak{Q}_1 \cap \mathfrak{Q}_2) = \mathfrak{P}$. Now let a be a zero-divisor modulo $\mathfrak{Q}_1 \cap \mathfrak{Q}_2$. Then we have a $b \not\equiv 0 \pmod{\mathfrak{Q}_1 \cap \mathfrak{Q}_2}$ such that $ab \equiv 0 \pmod{\mathfrak{Q}_1 \cap \mathfrak{Q}_2}$. Since $b \not\equiv 0 \pmod{\mathfrak{Q}_1 \cap \mathfrak{Q}_2}$ we can suppose that $b \not\equiv 0 \pmod{\mathfrak{Q}_1}$. Then $ab \equiv 0 \pmod{\mathfrak{Q}_1}$, gives $a \equiv 0 \pmod{\mathfrak{R}(\mathfrak{Q}_1)}$. Hence $a \, \varepsilon \, \mathfrak{P}$.

We can use this result to combine primary factors that have the same associated primes. In this way we obtain a representation of $\mathfrak{B}$ as irredundant intersection of primary ideals:

$$(11) \qquad \mathfrak{B} = \mathfrak{Q}_1 \cap \mathfrak{Q}_2 \cap \cdots \cap \mathfrak{Q}_r$$

such that the associated prime ideals $\mathfrak{P}_1, \mathfrak{P}_2, \cdots, \mathfrak{P}_r$ are distinct. Even after these normalizations have been made we cannot assert that the $\mathfrak{Q}_i$ are unique. For example, in $\mathfrak{F}[x,y]$ we have the distinct decompositions

$$(x^2, xy) = (x) \cap (x^2, xy, y^2)$$

$$= (x) \cap (x^2, y + \alpha x), \quad \alpha \, \varepsilon \, \mathfrak{F}.$$

We note, however, that the associated prime ideals of these two decompositions, namely, (x) and (x,y) are the same and this unicity carries over in general. This is the content of the

First uniqueness theorem. *Let* $\mathfrak{B} = \mathfrak{Q}_1 \cap \mathfrak{Q}_2 \cap \cdots \cap \mathfrak{Q}_r = \mathfrak{Q}_1' \cap \mathfrak{Q}_2' \cap \cdots \cap \mathfrak{Q}_s'$ *be two irredundant intersections into primary ideals whose associated primes are distinct. Then* $r = s$ *and the sets of primes of the two decompositions are identical.*

Before proceeding to the proof we shall derive a couple of simple lemmas.

Lemma 2. *Let* $\mathfrak{Q}$ *be a primary ideal and let* $\mathfrak{P}'$ *be a prime ideal containing* $\mathfrak{Q}$. *Then* $\mathfrak{P}' \supseteq \mathfrak{P} = \mathfrak{R}(\mathfrak{Q})$.

Proof. If $z \equiv 0 \pmod{\mathfrak{P}}$, $z^r \equiv 0 \pmod{\mathfrak{Q}}$ for some integer r. Hence $z^r \equiv 0 \pmod{\mathfrak{P}'}$. Since $\mathfrak{P}'$ is prime, $z \equiv 0 \pmod{\mathfrak{P}'}$.

Lemma 3. *Let* $\mathfrak{Q}$ *be primary,* $\mathfrak{P}$ *its associated prime and let* $\mathfrak{C}$ *be any ideal not contained in* $\mathfrak{P}$; *then* $\mathfrak{Q}:\mathfrak{C} = \mathfrak{Q}$.

Proof. An element u in $\mathfrak{Q}:\mathfrak{C}$ satisfies the condition that $uc \equiv 0 \pmod{\mathfrak{Q}}$ for all $c \in \mathfrak{C}$. If we choose $c \not\equiv 0 \pmod{\mathfrak{P}}$, then this implies that $u \equiv 0 \pmod{\mathfrak{Q}}$. Hence $\mathfrak{Q}:\mathfrak{C} \subseteq \mathfrak{Q}$. The converse $\mathfrak{Q} \subseteq \mathfrak{Q}:\mathfrak{C}$ is clear.

We can now give the

Proof of the uniqueness theorem. Let $\mathfrak{P}_i = \mathfrak{R}(\mathfrak{Q}_i)$, $\mathfrak{P}_i' = \mathfrak{R}(\mathfrak{Q}_i')$. There exist ideals in the set $\mathfrak{P}_1, \mathfrak{P}_2, \cdots, \mathfrak{P}_r, \mathfrak{P}_1', \mathfrak{P}_2', \cdots, \mathfrak{P}_s'$ that are not contained properly in any of the ideals of this collection. We may suppose that $\mathfrak{P}_1$ has this property. We prove first that $\mathfrak{P}_1$ is also in the set $\mathfrak{P}_1', \mathfrak{P}_2', \cdots, \mathfrak{P}_s'$. If not, then $\mathfrak{P}_1 \not\subseteq \mathfrak{P}_i'$ for $i = 1, 2, \cdots, s$. Hence, by Lemma 2, $\mathfrak{Q}_1 \not\subseteq \mathfrak{P}_i'$. By Lemma 3, $\mathfrak{Q}_i':\mathfrak{Q}_1 = \mathfrak{Q}_i'$. Hence

$$\mathfrak{B}:\mathfrak{Q}_1 = (\mathfrak{Q}_1' \cap \mathfrak{Q}_2' \cap \cdots \cap \mathfrak{Q}_s'):\mathfrak{Q}_1$$

$$= \mathfrak{Q}_1':\mathfrak{Q}_1 \cap \mathfrak{Q}_2':\mathfrak{Q}_1 \cap \cdots \cap \mathfrak{Q}_s':\mathfrak{Q}_1$$

$$= \mathfrak{Q}_1' \cap \mathfrak{Q}_2' \cap \cdots \cap \mathfrak{Q}_s' = \mathfrak{B}.$$

Similarly, $\mathfrak{Q}_j:\mathfrak{Q}_1 = \mathfrak{Q}_j$ if $j > 1$. Hence

$$\mathfrak{B} = \mathfrak{B}:\mathfrak{Q}_1 = (\mathfrak{Q}_1 \cap \mathfrak{Q}_2 \cap \cdots \cap \mathfrak{Q}_r):\mathfrak{Q}_1 = \mathfrak{Q}_2 \cap \mathfrak{Q}_3 \cap \cdots \cap \mathfrak{Q}_r$$

and this contradicts the assumption that the first decomposition is irredundant.

We now suppose that $\mathfrak{P}_1 = \mathfrak{P}_1'$. The ideal $\mathfrak{Q}_1 \cap \mathfrak{Q}_1'$ is primary with $\mathfrak{P}_1$ as associated prime. Hence, by the argument that we have just used, $\mathfrak{Q}_j:(\mathfrak{Q}_1 \cap \mathfrak{Q}_1') = \mathfrak{Q}_j$ for $j > 1$ and $\mathfrak{Q}_i':(\mathfrak{Q}_1 \cap \mathfrak{Q}_1') = \mathfrak{Q}_i'$ for $i > 1$. Hence

$$\mathfrak{B}:(\mathfrak{Q}_1 \cap \mathfrak{Q}') = \mathfrak{Q}_2 \cap \mathfrak{Q}_3 \cap \cdots \cap \mathfrak{Q}_r$$

$$= \mathfrak{Q}_2' \cap \mathfrak{Q}_3' \cap \cdots \cap \mathfrak{Q}_s'$$

and these are two irredundant decompositions of $\mathfrak{B}:(\mathfrak{Q}_1 \cap \mathfrak{Q}_2)$ satisfying the conditions of the theorem. We can use induction to conclude that the sets of prime ideals $\mathfrak{P}_2, \mathfrak{P}_3, \cdots, \mathfrak{P}_r$ coincides with the set $\mathfrak{P}_2', \mathfrak{P}_3', \cdots, \mathfrak{P}_s'$. This concludes the proof.

We shall call the prime ideals $\mathfrak{P}_1, \mathfrak{P}_2, \cdots, \mathfrak{P}_r$ whose uniqueness has just been established the *associated primes* of the ideal $\mathfrak{B}$. If $\mathfrak{B} = \mathfrak{Q}_1'' \cap \mathfrak{Q}_2'' \cap \cdots \cap \mathfrak{Q}_t''$ is any irredundant decomposition of $\mathfrak{B}$ into primary ideals, we can obtain a decomposition of the type considered in the theorem by combining components that have the same associated primes. Hence the distinct associated primes of the primary ideals $\mathfrak{Q}_1'', \mathfrak{Q}_2'', \cdots, \mathfrak{Q}_t''$ are the associated primes of $\mathfrak{B}$.

It is an immediate corollary of the uniqueness theorem that $\mathfrak{B}$ is primary if and only if it has only one associated prime. In other words, an ideal that is an irredundant intersection of primary ideals that do not all have the same associated prime is not primary.

Before proceeding to the discussion of the next uniqueness theorem we prove the following important

Theorem 6. *If $\mathfrak{B}$ and $\mathfrak{C}$ are ideals in a Noetherian ring, $\mathfrak{B}:\mathfrak{C} = \mathfrak{B}$ if and only if $\mathfrak{C}$ is not contained in any of the associated primes of $\mathfrak{B}$.*

Proof. Let $\mathfrak{B} = \mathfrak{Q}_1 \cap \mathfrak{Q}_2 \cap \cdots \cap \mathfrak{Q}_r$ be an irredundant decomposition of $\mathfrak{B}$ into primary ideals. Let $\mathfrak{P}_i = \mathfrak{R}(\mathfrak{Q}_i)$ and assume that $\mathfrak{C} \not\subseteq \mathfrak{P}_i$. Then by Lemma 3, $\mathfrak{Q}_i:\mathfrak{C} = \mathfrak{Q}_i$. Hence

$$\mathfrak{B}:\mathfrak{C} = (\mathfrak{Q}_1 \cap \mathfrak{Q}_2 \cap \cdots \cap \mathfrak{Q}_r):\mathfrak{C}$$

$$= \mathfrak{Q}_1:\mathfrak{C} \cap \mathfrak{Q}_2:\mathfrak{C} \cap \cdots \cap \mathfrak{Q}_r:\mathfrak{C}$$

$$= \mathfrak{Q}_1 \cap \mathfrak{Q}_2 \cap \cdots \cap \mathfrak{Q}_r = \mathfrak{B}.$$

On the other hand, suppose that $\mathfrak{C} \subseteq \mathfrak{P}_i$ for some i, say, $\mathfrak{C} \subseteq \mathfrak{P}_1$. Then there exists an integer m such that $\mathfrak{C}^m \subseteq \mathfrak{Q}_1$. Hence

$$\mathfrak{C}^m(\mathfrak{Q}_2 \cap \cdots \cap \mathfrak{Q}_r) \subseteq \mathfrak{C}^m \cap \mathfrak{Q}_2 \cap \cdots \cap \mathfrak{Q}_r \subseteq \mathfrak{B}.$$

Now let n be the smallest integer such that

$$(12) \qquad\qquad \mathfrak{C}^n(\mathfrak{Q}_2 \cap \cdots \cap \mathfrak{Q}_r) \subseteq \mathfrak{B}.$$

Since $\mathfrak{B} = \mathfrak{Q}_1 \cap \cdots \cap \mathfrak{Q}_r$ is irredundant, $n \geq 1$. It follows that $\mathfrak{C}^{n-1}(\mathfrak{Q}_2 \cap \cdots \cap \mathfrak{Q}_r) \not\subseteq \mathfrak{B}$.*   On the other hand, by (12) $\mathfrak{C}^{n-1}(\mathfrak{Q}_2 \cap \cdots \cap \mathfrak{Q}_r) \subseteq \mathfrak{B}:\mathfrak{C}$.   Hence $\mathfrak{B}:\mathfrak{C} \supset \mathfrak{B}$.

Suppose now that we also have a decomposition of $\mathfrak{C}$ as an irredundant intersection $\mathfrak{Q}_1' \cap \mathfrak{Q}_2' \cap \cdots \cap \mathfrak{Q}_s'$ with associated primes $\mathfrak{P}_1', \mathfrak{P}_2', \cdots, \mathfrak{P}_s'$. Then if $\mathfrak{C} \subseteq \mathfrak{P}_1$, $\mathfrak{Q}_1'\mathfrak{Q}_2' \cdots \mathfrak{Q}_s' \subseteq \mathfrak{P}_1$. Hence one of the $\mathfrak{Q}_j'$ and consequently one of the $\mathfrak{P}_j'$ is contained in $\mathfrak{P}_1$.   Conversely, it is clear that, if $\mathfrak{P}_j' \subseteq \mathfrak{P}_1$, then $\mathfrak{C} \subseteq \mathfrak{P}_j' \subseteq \mathfrak{P}_1$. Using this remark, we can reformulate the criterion that we have just derived as follows:

Theorem 6'.   *If $\mathfrak{B}$ and $\mathfrak{C}$ are ideals in a Noetherian ring, then $\mathfrak{B}:\mathfrak{C} = \mathfrak{B}$ if and only if no associated prime of $\mathfrak{C}$ is contained in any of the associated primes of $\mathfrak{B}$.*

We shall now use this criterion to derive the second uniqueness theorem. This concerns the isolated components of an ideal $\mathfrak{B}$. If $\mathfrak{B}$ is represented as an irredundant intersection $\mathfrak{Q}_1 \cap \mathfrak{Q}_2 \cap \cdots \cap \mathfrak{Q}_r$ where the $\mathfrak{Q}_i$ are primary and have distinct primes $\mathfrak{P}_1, \mathfrak{P}_2, \cdots, \mathfrak{P}_r$, then a particular $\mathfrak{Q}$ is called an *isolated primary component* of $\mathfrak{B}$ if the prime associated with $\mathfrak{Q}$ contains no other associated prime of $\mathfrak{B}$. More generally we call $\mathfrak{Q}_{i_1} \cap \mathfrak{Q}_{i_2} \cap \cdots \cap \mathfrak{Q}_{i_n}$ an *isolated component* of $\mathfrak{B}$ if no $\mathfrak{P}_{i_j}$ associated with the displayed primary ideals contains any of the associated primes that are not in this set. We can now state the

Second uniqueness theorem.   *Let $\mathfrak{B} = \mathfrak{Q}_1 \cap \mathfrak{Q}_2 \cap \cdots \cap \mathfrak{Q}_r = \mathfrak{Q}_1' \cap \mathfrak{Q}_2' \cap \cdots \cap \mathfrak{Q}_r'$ be two decompositions of $\mathfrak{B}$ that satisfy the conditions of the first uniqueness theorem. Let $\mathfrak{C} = \mathfrak{Q}_{i_1} \cap \mathfrak{Q}_{i_2} \cap \cdots \cap \mathfrak{Q}_{i_k}$ be an isolated component in the first decomposition and let $\mathfrak{C}'$ be the isolated component of the second decomposition that has the same set of associated primes as $\mathfrak{C}$. Then $\mathfrak{C} = \mathfrak{C}'$.*

---

* We use the convention that $\mathfrak{C}^0(\mathfrak{Q}_2 \cap \cdots \cap \mathfrak{Q}_r) = \mathfrak{Q}_2 \cap \cdots \cap \mathfrak{Q}_r$.

Proof. Write $\mathfrak{B} = \mathfrak{C} \cap \mathfrak{D} = \mathfrak{C}' \cap \mathfrak{D}'$ where $\mathfrak{D}$ and $\mathfrak{D}'$ are respectively the intersections of the $\mathfrak{Q}_i$ and the $\mathfrak{Q}_i'$ that do not contain $\mathfrak{C}$ and $\mathfrak{C}'$. Then the associated primes of $\mathfrak{D} \cap \mathfrak{D}'$ are contained in none of the associated primes of $\mathfrak{C}$. Hence $\mathfrak{C}:(\mathfrak{D} \cap \mathfrak{D}') = \mathfrak{C}$. Similarly, $\mathfrak{C}':(\mathfrak{D} \cap \mathfrak{D}') = \mathfrak{C}'$. Hence

$$\mathfrak{B}:(\mathfrak{D} \cap \mathfrak{D}') = (\mathfrak{C}:(\mathfrak{D} \cap \mathfrak{D}')) \cap (\mathfrak{D}:(\mathfrak{D} \cap \mathfrak{D}')) = \mathfrak{C}$$

and

$$\mathfrak{B}:(\mathfrak{D} \cap \mathfrak{D}') = (\mathfrak{C}':(\mathfrak{D} \cap \mathfrak{D}')) \cap (\mathfrak{D}':(\mathfrak{D} \cap \mathfrak{D}')) = \mathfrak{C}'.$$

Thus $\mathfrak{C} = \mathfrak{C}'$.

Note: Another uniqueness theorem, namely, the uniqueness of the number of irreducible components of an ideal will be proved in § **5** of the next chapter.

EXERCISES

1. Prove that, if all the associated prime ideals of $\mathfrak{B}$ are maximal, then there is only one decomposition of $\mathfrak{B}$ as an irredundant intersection of primary ideals with distinct associated primes.

2. Prove that the radical of an ideal in a Noetherian ring is the intersection of the associated prime ideals.

3. Prove that the radical is a prime ideal if and only if the given ideal has only one isolated primary component.

4. If $\mathfrak{B}$ is an ideal, we define the *ω-th power* of $\mathfrak{B}$, $\mathfrak{B}^\omega$, to be $\bigcap_i \mathfrak{B}^i, i = 1, 2,$ 3, $\cdots$. Let $\mathfrak{B}$ be an ideal in a Noetherian ring and write $\mathfrak{B}^\omega \mathfrak{B} = \mathfrak{Q}_1 \cap \mathfrak{Q}_2 \cap \cdots \cap \mathfrak{Q}_n$ an irredundant intersection of primary ideals. Prove that $\mathfrak{Q}_j \supseteq \mathfrak{B}^\omega$ for $j = 1, 2, \cdots, n$. Hence show that $\mathfrak{B}^\omega \mathfrak{B} = \mathfrak{B}^\omega$.

9. Integral dependence. The notion that we shall consider next is a generalization of the classical concept of an algebraic integer. A complex number is called an *algebraic integer* if it is a root of a polynomial with integer coefficients and leading coefficient 1. Now let $\mathfrak{A}$ be any commutative ring with an identity and let $\mathfrak{g}$ be a subring of $\mathfrak{A}$ containing 1. Then we shall say that an element $a \, \varepsilon \, \mathfrak{A}$ is *integrally dependent on* $\mathfrak{g}$ or is a *$\mathfrak{g}$-integer* if a satisfies an equation $f(x) = 0$ where $f(x) \, \varepsilon \, \mathfrak{g}[x]$ and has leading coefficient 1. If we write $f(x) = x^n - \gamma_1 x^{n-1} - \cdots - \gamma_{n-1}, \gamma_i$ in $\mathfrak{g}$, then we have

$$(13) \qquad\qquad a^n = \gamma_0 + \gamma_1 a + \cdots + \gamma_{n-1} a^{n-1}.$$

It follows from this that all the powers of a are expressible as linear combinations of $1, a, \cdots, a^{n-1}$ using coefficients in $\mathfrak{g}$.

Now we regard $\mathfrak{A}$ as a $\mathfrak{g}$-module in the obvious way: the group of the module is $\mathfrak{A}, +$, and multiplication by elements of $\mathfrak{g}$ is ring multiplication. Then the result that we have observed is that if a is a $\mathfrak{g}$-integer and (13) holds, then all the powers of a are contained in the finitely generated $\mathfrak{g}$-module $(1, a, \cdots, a^{n-1})$. The converse is clear; for, if $a^n \,\varepsilon\, (1, a, \cdots, a^{n-1})$, then we have a relation of the form (13).

In the remainder of this section we shall assume that $\mathfrak{g}$ is Noetherian and we shall investigate the totality of $\mathfrak{g}$-integral elements. The main tool in our considerations will be the following module criterion

Theorem 7. *If $\mathfrak{g}$ is Noetherian, an element $a \,\varepsilon\, \mathfrak{A}$ is a $\mathfrak{g}$-integer if and only if there exists a finitely generated submodule of $\mathfrak{A}$ that contains all the powers of a.*

Proof. We have just seen that this condition is necessary. Now let $\mathfrak{N}$ be a finitely generated $\mathfrak{g}$-module containing all the powers of a. Since $\mathfrak{g}$ is Noetherian, $\mathfrak{N}$ satisfies the ascending chain condition for submodules. Hence, there exists an integer n such that in the ascending chain

$$(1) \subseteq (1,a) \subseteq (1,a,a^2) \subseteq \cdots$$

we have $(1, a, \cdots, a^{n-1}) = (1, a, \cdots, a^n)$. This implies that $a^n \,\varepsilon\, (1, a, \cdots, a^{n-1})$ so that we have a relation of the form (13).

We use this criterion to prove first the following

Theorem 8. *The totality $\mathfrak{G}$ of elements of $\mathfrak{A}$ that are $\mathfrak{g}$-integral is a subring of $\mathfrak{A}$ containing $\mathfrak{g}$.*

Proof. Any element γ of $\mathfrak{g}$ satisfies an equation $x - \gamma = 0$. Hence, it belongs to $\mathfrak{G}$. Next let a and $b \,\varepsilon\, \mathfrak{G}$ and let $(u_1, u_2, \cdots, u_s)$ and $(v_1, v_2, \cdots, v_t)$ be $\mathfrak{g}$-modules of $\mathfrak{A}$ that contain all the powers of a and of b respectively. The product of any element of (u_i) by any element of (v_j) is in the submodule

$$\mathfrak{P} = (u_1 v_1, \cdots, u_1 v_t; u_2 v_1, \cdots, u_2 v_t; \cdots; \cdots u_s v_t).$$

Hence, any monomial of the form $a^k b^l \,\varepsilon\, \mathfrak{P}$. It follows that all the powers of $a \pm b$ and of ab are in $\mathfrak{P}$. Hence, $a + b$ and $ab \ \varepsilon \ \mathfrak{G}$ and $\mathfrak{G}$ is a subring of $\mathfrak{A}$.

We shall say that $\mathfrak{g}$ is *integrally closed in* $\mathfrak{A}$ if $\mathfrak{G} = \mathfrak{g}$, that is, if every element of $\mathfrak{A}$ that is integrally dependent on $\mathfrak{g}$ belongs to $\mathfrak{g}$. We prove next

Theorem 9. *The ring $\mathfrak{G}$ of $\mathfrak{g}$-integral elements is integrally closed in $\mathfrak{A}$.*

Proof. Let a be a $\mathfrak{G}$-integer and let

$$a^n = g_0 + g_1 a + \cdots + g_{n-1} a^{n-1}$$

where the $g_i \ \varepsilon \ \mathfrak{G}$. We can use this relation to show that every power of a is expressible as a linear combination of the powers 1, $a, \cdots, a^{n-1}$ using coefficients that are sums of monomials in the g's. A simple extension of the argument used to prove the preceding theorem shows that there exists a finitely generated $\mathfrak{g}$-submodule $(w_1, w_2, \cdots, w_l)$ of $\mathfrak{A}$ that contains all the monomials in the g's. Then it is clear that every power of a is contained in

$$(w_1, \cdots, w_l; w_1 a, \cdots, w_l a; \cdots; \cdots w_l a^{n-1}).$$

Hence $a \ \varepsilon \ \mathfrak{G}$ as we wished to show.

If $\mathfrak{A} = \mathfrak{F}$ is a field and $\mathfrak{g} = \mathfrak{F}_0$ is a subfield, then an element of $\mathfrak{F}$ is $\mathfrak{F}_0$-integral if and only if it is algebraic over $\mathfrak{F}_0$ (§ 7, p. 100). Hence, Theorem 8 states in this case that the set $\mathfrak{G}$ of elements of $\mathfrak{F}$ that are algebraic over $\mathfrak{F}_0$ is a subring of $\mathfrak{F}$ containing $\mathfrak{F}_0$. Also we know that, if a is algebraic, then $\mathfrak{F}_0[a]$ is a subfield. Hence, if $a \neq 0$, $a^{-1} \ \varepsilon \ \mathfrak{F}_0[a] \subseteq \mathfrak{G}$. Hence, $\mathfrak{G}$ is a field. If we take into account also Theorem **9**, we can state the following important theorem on fields.

Theorem 10. *Let $\mathfrak{F}$ be a field and $\mathfrak{F}_0$ a subfield. Then the set $\mathfrak{G}$ of elements of $\mathfrak{F}$ that are algebraic over $\mathfrak{F}_0$ forms a subfield of $\mathfrak{F}$ containing $\mathfrak{F}_0$. Any element of $\mathfrak{F}$ that is algebraic over $\mathfrak{G}$ belongs to $\mathfrak{G}$.*

Now let $\mathfrak{F}$ be any field, let $\mathfrak{g}$ be any subring of $\mathfrak{F}$ containing 1 and let $\mathfrak{F}_0$ denote the subfield of $\mathfrak{F}$ generated by $\mathfrak{g}$. If an element $a \ \varepsilon \ \mathfrak{F}$ is $\mathfrak{g}$-integral, it is certainly algebraic over $\mathfrak{F}_0$. Hence, its minimum polynomial $\mu(x)$ has coefficients in $\mathfrak{F}_0$ and leading coefficient 1. We shall now show that, if $\mathfrak{g}$ is Gaussian, $\mu(x) \ \varepsilon \ \mathfrak{g}[x]$.

To see this, let $f(x)$ be some polynomial with leading coefficient 1 and other coefficients in $\mathfrak{g}$ such that $f(a) = 0$. Then $\mu(x) \mid f(x)$. Now one of the irreducible factors of $f(x)$ in $\mathfrak{g}[x]$ is an associate of $\mu(x)$ in $\mathfrak{F}_0[x]$. If we call this factor $\mu^*(x)$, then $\mu^*(x) = \beta\mu(x)$, β in $\mathfrak{F}_0$. Since the leading coefficient of $f(x)$ is 1 and $\mu^*(x) \mid f(x)$, we can suppose that the leading coefficient of $\mu^*(x)$ is 1. Then the relation $\mu^*(x) = \beta\mu(x)$ gives $\beta = 1$ so that $\mu(x) = \mu^*(x) \in \mathfrak{g}[x]$. This proves the following

Theorem 11. *Let $\mathfrak{g}$ be a Gaussian subring of a field $\mathfrak{F}$ and let $\mathfrak{F}_0$ be the subfield of $\mathfrak{F}$ generated by $\mathfrak{g}$. Then an element $a \in \mathfrak{F}$ is integrally dependent on $\mathfrak{g}$ if and only if it is algebraic over $\mathfrak{F}_0$ and its minimum polynomial over $\mathfrak{F}_0$ has coefficients in $\mathfrak{g}$.*

This criterion is particularly useful if every element of $\mathfrak{F}$ is algebraic over $\mathfrak{F}_0$; for in this case it asserts that an element of $\mathfrak{F}$ is $\mathfrak{g}$-integral if and only if its minimum polynomial is in $\mathfrak{g}[x]$. We note also that, since the elements of $\mathfrak{F}_0$ are algebraic over $\mathfrak{F}_0$ and have minimum polynomials of the form $x - \gamma$, the only elements of $\mathfrak{F}_0$ that are integral over $\mathfrak{g}$ are those in $\mathfrak{g}$. Then $\mathfrak{g}$ is integrally closed in $\mathfrak{F}_0$. An integral domain is said to be *integrally closed* if it is integrally closed in its field of fractions. The result that we have obtained can therefore be stated as the following

Corollary. *Any Gaussian integral domain is integrally closed.*

10. Integers of quadratic fields. The theory of algebraic numbers is concerned with the arithmetic properties of fields of the form $R_0(\theta)$ where R_0 is the field of rational numbers and θ is an algebraic element. The primary object of study in this theory is the ring $\mathfrak{G}$ of elements of $R_0(\theta)$ that are I-integers (or simply *integers* of $R_0(\theta)$). In this section we give a brief introduction to the theory of algebraic numbers by determining the ring of integers of quadratic extensions $R_0(\theta)$.

Let m be an (ordinary) integer that has no square factors. Then the polynomial $x^2 - m$ is irreducible in $I[x]$. Since I is Gaussian, it follows that $x^2 - m$ is irreducible in $R_0[x]$. Hence, we can construct an extension field $R_0(\theta)$ where $\theta^2 = m$. Such a field is called a *quadratic extension* of the field of rational numbers.

Any element of $R_0(\theta)$ can be written in one and only one way in the form $u = \alpha + \beta\theta$ where α and $\beta \, \varepsilon \, R_0$. If $u = \alpha + \beta\theta$, we define the *conjugate* (in $R_0(\theta)$) of u to be the element $\bar{u} = \alpha - \beta\theta$ of this field. It is easy to verify that the mapping $u \to \bar{u}$ is an automorphism of $R_0(\theta)$. Also it is clear that, if u is not in R_0, then $\bar{u} \neq u$. We set

$$T(u) = u + \bar{u} = 2\alpha, \quad N(u) = u\bar{u} = \alpha^2 - \beta^2 m,$$

and note that $T(u)$ and $N(u)$ are in R_0. Hence, the polynomial

$$f(x,u) = (x - u)(x - \bar{u}) = x^2 - T(u)x + N(u)$$

has rational coefficients. Evidently u is a root of $f(x,u)$. Hence, every element of $R_0(\theta)$ is algebraic over R_0.

If $u \, \varepsilon \, R_0$, u is integrally dependent on I if and only if it belongs to I. If $u \notin R_0$, then the minimum polynomial of u relative to R_0 is of degree >1. Hence, it is the polynomial $f(x,u)$. Then u is an integer of $R_0(\theta)$ if and only if the coefficients $T(u)$ and $N(u)$ are integers. Thus we have the conditions

$$(14) \qquad\qquad 2\alpha \, \varepsilon \, I, \quad \alpha^2 - \beta^2 m \, \varepsilon \, I.$$

The first of these conditions implies that either $\alpha \, \varepsilon \, I$ or that α is half of an odd integer, say, $\alpha = (2n + 1)/2$. If $\alpha \, \varepsilon \, I$ the second condition gives $\beta^2 m \, \varepsilon \, I$. Since m has no square factors, this implies that $\beta \, \varepsilon \, I$; for otherwise $\beta = b_1 b_2^{-1}$ where b_1 and $b_2 \, \varepsilon \, I$ and b_2 is divisible by a prime p that does not divide b_1. Then

$$b_1^2 m = (\beta^2 m) b_2^2 \equiv 0 \quad (\bmod \ p^2).$$

Since $p \nmid b_1$, this implies that $p^2 \mid m$ contrary to our assumption.

Suppose next that $\alpha = (2n + 1)/2$, n in I. In this case the condition that $N \equiv \alpha^2 - \beta^2 m \, \varepsilon \, I$ gives

$$\beta^2 m = \alpha^2 - N = (4n^2 + 4n - 4N + 1)/4.$$

Hence

$$(15) \qquad\qquad \beta^2 m = (4r + 1)/4, \quad r \, \varepsilon \, I.$$

Now write $\beta = b_1 b_2^{-1}$ where b_1 and b_2 are integers such that $(b_1, b_2) = 1$ and multiply (15) by $4b_2^2$. This gives

$$4b_1^2 m = (4r + 1)b_2^2.$$

Since m is square-free and $(b_1, b_2) = 1$, this relation implies that $b_2{}^2 = 4$ and $b_2 = \pm 2$. Thus b_1 is odd and β is half of an odd integer.

Now write $\beta = (2q + 1)/2$ as well as $\alpha = (2n + 1)/2$. Since

$$N = \alpha^2 - \beta^2 m = [4n^2 + 4n + 1 - (4q^2 + 4q + 1)m]/4$$

is an integer, we have the congruence

$$4n^2 + 4n + 1 - (4q^2 + 4q + 1)m \equiv 0 \quad (\text{mod } 4).$$

This reduces to $1 - m \equiv 0$ (mod 4) and $m \equiv 1$ (mod 4). Thus we see that, unless m is of the form $4k + 1$, the integers of $R_0(\theta)$, $\theta^2 = m$, are necessarily of the form $\alpha + \beta\theta$ where α and β are ordinary integers. If $m \equiv 1$ (mod 4), then we also have the possibility that an integer has the form $\alpha + \beta\theta$ where α and β are both halves of odd integers.

Conversely, if α and $\beta \, \varepsilon \, I$, then (13) holds and $\alpha + \beta\theta$ is a quadratic integer. Also, if $m \equiv 1$ (mod 4) and α and β are halves of odd integers, then $\alpha + \beta\theta$ is a quadratic integer. Our conclusions can be summarized as follows:

Theorem 12. *If m is a square free integer $\equiv 2$ or 3 (mod 4), then the ring $\circledS$ of integers of $R_0(\theta)$ is the set of numbers of the form $\alpha + \beta\theta$ where α and $\beta \, \varepsilon \, I$. If $m \equiv 1$ (mod 4), $\circledS$ is the set of numbers of the form $\alpha + \beta\theta$ where α and β are either both in I or both halves of odd integers.*

EXERCISES

1. Show that if $m = -3$, $\circledS$ is Euclidean.

2. Prove that there are just five negative values of m, namely, $m = -1$, -2, -3, -7, -11 such that $\circledS$ is Euclidean relative to the function $\delta(\alpha) = |N(\alpha)|$.[*]

---

[*] See for example Hardy and Wright, *The Theory of Numbers*, Oxford, 1938, p. 213. The positive values of m for which this holds have been determined only recently. They are $m = 2, 3, 5, 6, 7, 11, 13, 17, 19, 21, 29, 33, 37, 41, 57, 73, 97$. See H. Chatland, *On the Euclidean algorithm in quadratic number fields*, Bull. Amer. Math. Soc., Vol. 55 (1949), pp. 948–953. The question of the existence of a Euclidean division process that does not necessarily make use of the function $\delta(\alpha) = |N(\alpha)|$ is discussed by T. Motzkin, in a paper, *The Euclidean algorithm*, Bull. Amer. Math. Soc., Vol. 55 (1949), pp. 1142–1146.

Chapter *VII*

LATTICES

In a number of important considerations in the theory of groups and of rings one is concerned primarily with certain distinguished subsets (invariant subgroups, ideals) of these systems rather than with the elements themselves. This is particularly true of the Jordan-Hölder-Schreier theory. Here the arguments concern the system of M-subgroups and the compositions in this system of intersection and group generated. Similarly, parts of the theory of rings are concerned with the systems of ideals (left, right, two-sided) of a ring and the compositions of intersection and sum in these systems. One is therefore led to the definition of an abstract system—called a lattice—that includes these two as instances. The concept of a lattice was first defined by Dedekind, but it attracted very little attention until quite recently (around 1930). Besides the applications to algebra many applications to the foundations of geometry and to other fields have been discovered. It should be noted also that prior to Dedekind's work a special class of lattices, Boolean algebras, had been introduced by Boole.

In this chapter we shall give a brief treatment of the parts of the theory of lattices that are applicable to group theory and ring theory. The arguments that we shall use will often be repetitions of those that we have encountered before. In such cases full details will be omitted.

1. Partially ordered sets

Definition 1. *A* partially ordered set *is a system consisting of a set S and a relation $\geq$ ("greater than or equals" or "contains") satisfying the following postulates:*

$\mathbf{P_1}$ *$a \geq b$ and $b \geq a$ hold if and only if $a = b$.*

$\mathbf{P_2}$ *If $a \geq b$ and $b \geq c$, then $a \geq c$.*

If a and b are any elements of S we may have $a \geq b$ or not; in the latter case we write $a \ngeq b$. Also if $a \geq b$ and $a \neq b$, then we write $a > b$, and we agree to use $b \leq a$ and $b < a$ as alternatives for $a \geq b$ and $a > b$.

Examples. (1) The set I of integers, the set P of positive integers and the set R of real numbers are partially ordered sets relative to the usual $\geq$ relation. (2) The set P of positive integers, the relation $\geq$ defined by the rule that $a \geq b$ if $a \mid b$. It is clear that $\mathbf{P_1}$ and $\mathbf{P_2}$ are satisfied. (3) The set $\mathfrak{P}$ of subsets of an arbitrary set S with $A \geq B$ defined to mean that B is a subset of A. (4) The set $\mathfrak{L}$ of subgroups of a group $\mathfrak{G}$ with $\mathfrak{H}_1 \geq \mathfrak{H}_2$ defined as in (3).

In any one of the examples, (2), (3), or (4), there exist elements a and b that are not comparable in the sense that neither $a \geq b$ nor $b \geq a$ holds. If every pair of elements of a partially ordered set S is comparable ($a \geq b$ or $b \geq a$), then S is said to be *linearly ordered* or is a *chain*. All of the examples in (1) are of this type.

In a finite partially ordered set the relation $>$ can be expressed in terms of the relation of covering. We say that a_1 is a *cover* of a_2 if $a_1 > a_2$ and no u exists such that $a_1 > u > a_2$. It is clear that, if $a > b$ in a finite partially ordered set, then we can find a chain

$$a = a_1 > a_2 > \cdots > a_n = b$$

in which each a_i covers a_{i+1}. Conversely the existence of such a chain implies that $a > b$. This remark enables us to represent any finite partially ordered set by a diagram. One obtains such a diagram by representing the elements of S by small circles (or dots) and placing the circle for a_1 above that for a_2 and connecting by a line if a_1 is a cover of a_2. Then $a > b$ if and only if there is a descending broken line connecting a to b. Some examples of such diagrams are the following:

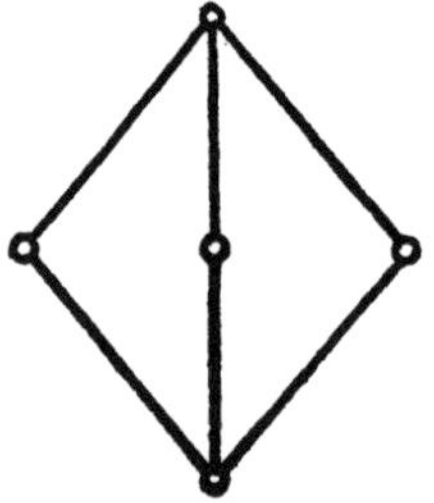 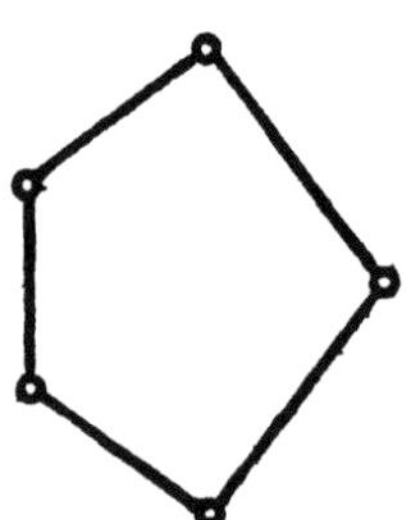 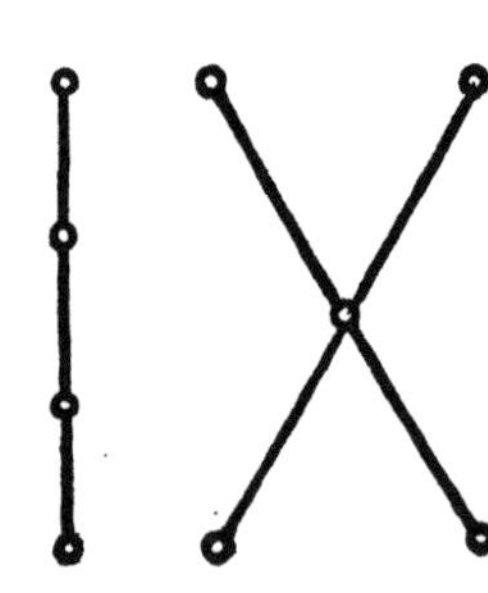

Evidently the notion of a diagram of a partially ordered set gives us another means to construct examples of such sets.

EXERCISES

1. Show that the partially ordered set of subgroups of a cyclic group of prime power order is a chain.

2. Let S be the set of all functions which are continuous over the interval $0 \leq x \leq 1$. Define $f \geq g$ if and only if $f(x) \geq g(x)$ for all x in the closed interval. Show that the relation $\geq$ is a partial ordering of S.

3. Obtain diagrams for the following partially ordered sets: the set of subsets of a set of three elements, the set of subgroups of the cyclic group of order 6, the set of subgroups of S_3.

2. Lattices. An element u of a partially ordered set S is said to be an upper bound for the subset A of S if $u \geq a$ for every $a \, \varepsilon \, A$. The element u is a *least upper bound* (l.u.b.) if u is an upper bound and $u \leq v$ for any upper bound v of A. It is immediate that if a least upper bound exists then it is unique. Similar definitions and remarks apply to lower bounds. These notions are fundamental in the following

Definition 2. *A* lattice (structure) *is a partially ordered set in which any two elements have a least upper bound and a greatest lower bound (g.l.b.).*

We denote the l.u.b. of a and b by $a \cup b$ ("a cup b" or "a union b") and the g.l.b. by $a \cap b$ ("a cap b" or "a intersect b"). If a,b,c are any three elements of a lattice L, then $(a \cup b) \cup c \geq a,b,c$. Moreover, if v is any element such that $v \geq a,b,c$ then $v \geq (a \cup b), c$. Hence $v \geq (a \cup b) \cup c$. Thus $(a \cup b) \cup c$ is a l.u.b. for a,b and c. A simple inductive argument shows that any finite subset of L has a l.u.b. Similarly any finite subset has a g.l.b. If the set consists of $a_1, a_2, \cdots, a_n$, then we denote these elements by

$$a_1 \cup a_2 \cup \cdots \cup a_n \quad \text{and} \quad a_1 \cap a_2 \cap \cdots \cap a_n$$

respectively.

A lattice L is said to be *complete* if any (finite or infinite) subset $A = \{a_\alpha\}$ has a l.u.b. $\cup a_\alpha$ and a g.l.b. $\cap a_\alpha$.

The examples (1)–(4) of partially ordered sets listed in § 1 are lattices. In the example (3) of subsets of a set, $A \cup B$ and $A \cap B$ have the usual significance of set-theoretic sum and set intersection. In the partially ordered set of subgroups of a group $\mathfrak{G}$, $\mathfrak{H}_1 \cup \mathfrak{H}_2$ is the group $[\mathfrak{H}_1, \mathfrak{H}_2]$ generated by $\mathfrak{H}_1$ and $\mathfrak{H}_2$ while $\mathfrak{H}_1 \cap \mathfrak{H}_2$ is the usual intersection. All of the diagrams given in § 1 except the last one represent lattices. The lattice of subsets of any set, and the lattice of subgroups of any group are complete. The lattice of rational numbers (the usual $\geq$) is not complete.

It is worth while to list the basic algebraic properties of the binary compositions $\cup$ and $\cap$ in a lattice. In doing so we shall be led to a second and somewhat more algebraic definition of a lattice.

We note first that the l.u.b. and the g.l.b. are symmetric functions of their arguments, that is, $a \cup b = b \cup a$ and $a \cap b = b \cap a$. Also we have seen that $(a \cup b) \cup c$ is the l.u.b. of a,b,c. Since the l.u.b. is unique,

$$(a \cup b) \cup c = (b \cup c) \cup a = a \cup (b \cup c).$$

Similarly

$$(a \cap b) \cap c = a \cap (b \cap c).$$

It is clear that

$$a \cup a = a, \quad a \cap a = a.$$

Since $a \cup b \geq a$, $(a \cup b) \cap a = a$. Similarly $(a \cap b) \cup a = a$.

Conversely suppose that L is any set in which there are defined two binary compositions $\cup$ and $\cap$ satisfying

$L_1 \qquad\qquad a \cup b = b \cup a, \quad a \cap b = b \cap a.$

$L_2 \quad (a \cup b) \cup c = a \cup (b \cup c), \quad (a \cap b) \cap c = a \cap (b \cap c).$

$L_3 \qquad\qquad a \cup a = a, \quad a \cap a = a.$

$L_4 \qquad (a \cup b) \cap a = a, \quad (a \cap b) \cup a = a.$

We shall show that L is a lattice relative to a suitable definition of $\geq$ and that $\cup$ and $\cap$ are the l.u.b. and the g.l.b. in this lattice.

Before proceeding to the proof we remark that we have made precisely the same assumptions on the two compositions $\cup$ and $\cap$. Hence, we have the important *principle of duality* that states that, if S is a statement which can be deduced from our axioms, then the *dual statement S'* obtained by interchanging $\cup$ and $\cap$ in S can also be deduced.

We note next that, if a and b belong to a system satisfying L_1–L_4, then the conditions $a \cup b = a$ and $a \cap b = b$ are equivalent; for, if $a \cup b = a$ holds, then $a \cap b = (a \cup b) \cap b = b$ and dually $a \cap b = b$ implies $a \cup b = a$. We shall now define a relation $\geq$ in L by specifying that $a \geq b$ means that either $a \cup b = a$ or $a \cap b = b$. Evidently in dualizing a statement $a \geq b$ has to be replaced by $b \geq a$.

We shall now show that the basic rules P_1–P_2 for partially ordered sets hold for the relation that we have introduced. Suppose that $a \geq b$ and $b \geq a$. Then $a \cup b = a$ and $b \cup a = b$. Hence by the commutative law $a = b$. Also by L_3 $a \cup a = a$ so that $a \geq a$. This proves P_1. Next assume that $a \geq b$ and $b \geq c$. Then $a \cup b = a$ and $b \cup c = b$. Hence,

$$a \cup c = (a \cup b) \cup c = a \cup (b \cup c) = a \cup b = a$$

and $a \geq c$. Hence P_2 holds.

Since $(a \cup b) \cap a = a$, $a \cup b \geq a$. Similarly $a \cup b \geq b$. Now let c be any element such that $c \geq a$ and $c \geq b$. Then $a \cup c = c$ and $b \cup c = c$. Hence

$$(a \cup b) \cup c = a \cup (b \cup c) = a \cup c = c$$

and $c \geq a \cup b$. This shows that $a \cup b$ is a l.u.b. of a and b. By duality $a \cap b$ is a g.l.b. of a and b. This concludes the proof that a system satisfying L_1–L_4 is a lattice.

A subset M of a lattice L is called a *sublattice* if it is closed relative to the compositions $\cup$ and $\cap$. It is evident that a sublattice is a lattice relative to the induced compositions. On the other hand, a subset of a lattice may be a lattice relative to the partial ordering $\geq$ defined in L without being a sublattice. For example, let $\mathfrak{G}$ be a group, let $\mathfrak{P}$ be the lattice of subsets of $\mathfrak{G}$, and $\mathfrak{L}$ be the lattice of subgroups of $\mathfrak{G}$. Then it is clear that $\mathfrak{L} \subseteq \mathfrak{P}$, and that $\mathfrak{H}_1 \geq \mathfrak{H}_2$ has the same significance in these two sets. On the other hand, if $\mathfrak{H}_1$ and $\mathfrak{H}_2$ are subgroups, then $\mathfrak{H}_1 \cup \mathfrak{H}_2$ as defined in $\mathfrak{P}$ is the set sum of these groups. In general, this is not a subgroup; hence, it differs from the $\mathfrak{H}_1 \cup \mathfrak{H}_2$ defined in $\mathfrak{L}$ as the smallest subgroup of $\mathfrak{G}$ containing $\mathfrak{H}_1$ and $\mathfrak{H}_2$.

If a is a fixed element of a lattice L, then the subset of elements x such that $x \geq a$ $(x \leq a)$ is evidently a sublattice. If $a \geq b$, the subset of elements x such that $a \geq x \geq b$ is a sublattice. We call such a sublattice a (closed) *interval* (*quotient*) and we denote it as $I[a,b]$.*

The definition of a lattice by means of the postulates L_1–L_4 leads also to the useful definition of homomorphism. A mapping $a \to a'$ of a lattice L into a lattice L' is called a *homomorphism* if $(a \cup b)' = a' \cup b'$ and $(a \cap b)' = a' \cap b'$. If such a mapping is 1–1, it is an *isomorphism*. A useful criterion for isomorphism is the following

Theorem 1. *A 1–1 mapping $a \to a'$ of a lattice L onto a lattice L' is an isomorphism if and only if $a \geq b$ in L implies and is implied by $a' \geq b'$ in L'.*

Proof. A mapping $a \to a'$ of a lattice L into a lattice L' is called *order preserving* if $a \geq b$ implies that $a' \geq b'$. If $a \to a'$ is an isomorphism and $a \geq b$, then $a \cup b = a$. Hence $a' \cup b' = a'$ and $a' \geq b'$. Thus $a \to a'$ is order preserving. Evidently the inverse mapping $a' \to a$ is also order preserving. Conversely, suppose that $a \to a'$ is a 1–1 mapping of L onto L' which is order preserving and whose inverse is also order preserving. Let $d = a \cup b$. Then $d \geq a,b$ so that $d' \geq a',b'$. Now let e' be any element of L' such that $e' \geq a',b'$ and let e be the element of L whose image is e'. Then $e \geq a,b$. Hence $e \geq d$ and $e' \geq d'$. This shows that $d' = a' \cup b'$. Similarly $(a \cap b)' = a' \cap b'$.

An element 1 of a partially ordered set is called an *all element* (*unit, identity*) if $1 \geq a$ for every a in the set. Dually, an element 0 is called a *zero element* if $0 \leq a$ for every a. Evidently, if these elements exist, they are unique.

EXERCISES

1. Show that the set of invariant subgroups and the set of M-subgroups (for any operator set M) are sublattices of the lattice of subgroups of any group.

2. Let S be the partially ordered set of ex. 2, p. 189. Define $f \cup g$ and $f \cap g$ suitably and prove that S forms a lattice with respect to these compositions and the given partial ordering. Is S a complete lattice?

3. Show that any complete lattice has a zero and an all element.

4. Prove that a partially ordered set with an all element in which every non-vacuous set has a g.l.b. is a complete lattice.

* This notation is more convenient for the algebraic applications than the usual one in which the smaller endpoint is displayed first.

3. Modular lattices. One of the compositions of a lattice, say $\cup$, can be regarded as the analogue of addition in a ring, while the other can be taken to be the analogue of multiplication. It is therefore natural to investigate lattices that are *distributive* in the sense that

$$(1) \qquad a \cap (b \cup c) = (a \cap b) \cup (a \cap c)$$

holds. Important examples of such lattices do exist. For instance, the lattice of all subsets of a set relative to the usual set theoretic sum and intersection is distributive. This is indicated in the figure

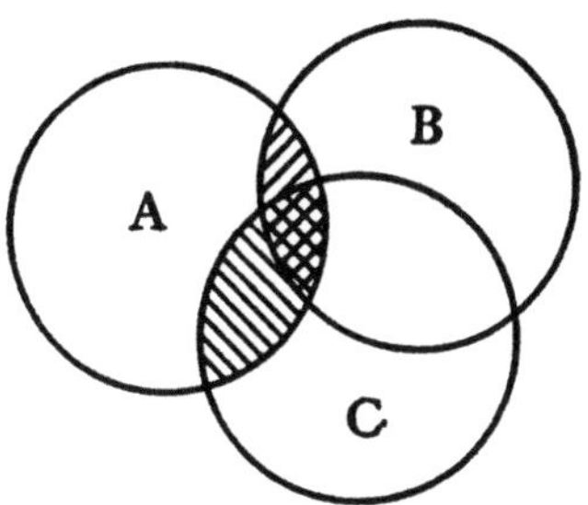

and is readily proved in general. Another example of a distributive lattice is the lattice of positive integers in which $a \geq b$ means that $a \mid b$. Here $a \cup b$ is the g.c.d. (a,b) and $a \cap b$ is the l.c.m. $[a,b]$ of a and b. Then (1) reads

$$[a,(b,c)] = ([a,b],[a,c]).$$

The proof of this follows easily from the properties of (a,b) and $[a,b]$ (ex. 2, p. 120).

It is clear that in any lattice $a \cap (b \cup c) \geq a \cap b$ and $a \cap (b \cup c) \geq a \cap c$. Hence

$$a \cap (b \cup c) \geq (a \cap b) \cup (a \cap c)$$

always holds. In order to establish distributivity it therefore suffices to prove the reverse inequality

$$a \cap (b \cup c) \leq (a \cap b) \cup (a \cap c).$$

We remark also that the condition (1) is equivalent to the dual condition:

$$(1') \qquad a \cup (b \cap c) = (a \cup b) \cap (a \cup c).$$

For if (1) holds, then

$$(a \cup b) \cap (a \cup c) = ((a \cup b) \cap a) \cup ((a \cup b) \cap c)$$
$$= a \cup ((a \cup b) \cap c)$$
$$= a \cup ((a \cap c) \cup (b \cap c))$$
$$= (a \cup (a \cap c)) \cup (b \cap c)$$
$$= a \cup (b \cap c).$$

Dually $(1')$ implies (1). Thus the assumption of (1) is equivalent to the assumption of (1) and $(1')$. Hence, it is clear that the principle of duality holds also for distributive lattices.

The most important lattices that occur in algebra (e.g., the lattices of ideals of rings) are not distributive. However, a number of these do satisfy a weaker form of (1) that reads as follows:

$\mathbf{L_5}$ If $a \geq b$, then $a \cap (b \cup c) = b \cup (a \cap c)$.

Since $b = a \cap b$ the right-hand side can be replaced by $(a \cap b) \cup (a \cap c)$. Thus our assumption amounts to the distributive law for triples a,b,c such that $a \geq b$. We now state the following important

Definition 3. *A lattice is called* modular (Dedekind) *if it satisfies the condition* L_5.

The importance of these lattices for the applications to other branches of algebra stems from the following

Theorem 2. *The lattice of invariant subgroups of any group is modular.*

Proof. Let $\mathfrak{G}$ be the given group and let $\mathfrak{H}_1, \mathfrak{H}_2, \mathfrak{H}_3$ be invariant subgroups such that $\mathfrak{H}_1 \geq \mathfrak{H}_2$ $(\mathfrak{H}_1 \supseteq \mathfrak{H}_2)$. Consider the intersection $\mathfrak{H}_1 \cap (\mathfrak{H}_2 \cup \mathfrak{H}_3)$ where $\mathfrak{H}_2 \cup \mathfrak{H}_3$ now denotes the l.u.b. of $\mathfrak{H}_2$ and $\mathfrak{H}_3$ in the lattice of subgroups. Thus $\mathfrak{H}_2 \cup \mathfrak{H}_3$ is the subgroup generated by $\mathfrak{H}_2$ and $\mathfrak{H}_3$. Since the $\mathfrak{H}_i$ are invariant, we know that $\mathfrak{H}_2 \cup \mathfrak{H}_3 = \mathfrak{H}_2\mathfrak{H}_3 = \mathfrak{H}_3\mathfrak{H}_2$. Hence, if $a \, \varepsilon \, \mathfrak{H}_1 \cap (\mathfrak{H}_2 \cup \mathfrak{H}_3)$, $a = h_1 \, \varepsilon \, \mathfrak{H}_1$ and $a = h_2h_3$ where $h_2 \, \varepsilon \, \mathfrak{H}_2$ and $h_3 \, \varepsilon \, \mathfrak{H}_3$. From $h_1 = h_2h_3$ we obtain $h_2^{-1}h_1 = h_3$. Since $\mathfrak{H}_1 \geq \mathfrak{H}_2$ the left-hand side of this equation represents an element of $\mathfrak{H}_1$. Hence $h_3 \, \varepsilon \, \mathfrak{H}_1$ and so $h_3 \, \varepsilon \, \mathfrak{H}_1 \cap \mathfrak{H}_3$. We have therefore proved the essential inequality

$$\mathfrak{H}_1 \cap (\mathfrak{H}_2 \cup \mathfrak{H}_3) \leq \mathfrak{H}_2 \cup (\mathfrak{H}_1 \cap \mathfrak{H}_3).$$

Previously we had noted that the reverse inequality is a general lattice theoretic property. Hence

$$\mathfrak{H}_1 \cap (\mathfrak{H}_2 \cup \mathfrak{H}_3) = \mathfrak{H}_2 \cup (\mathfrak{H}_1 \cap \mathfrak{H}_3),$$

and the theorem is proved.

It is clear that any sublattice of a modular lattice is modular. Hence the lattice of invariant M-subgroups of any M-group is modular. Hence, also the lattice of submodules of any module and the lattices of ideals (left, right, two-sided) of any ring are modular. On the other hand, the lattice of all subgroups of a group is generally not modular. This fact makes it somewhat unnatural to try to subsume all of group theory under the theory of lattices.*

We note that the principle of duality holds in modular lattices; for the dual of L_5 reads: if $a \leq b$, then $a \cup (b \cap c) = b \cap (a \cup c)$, and this clearly means the same thing as L_5. An alternative useful definition of a modular lattice can be extracted from the following

Theorem 3. *A lattice L is modular if and only if $a \geq b$ and $a \cup c = b \cup c$, $a \cap c = b \cap c$ for any c imply that $a = b$.*

Proof. Let L be modular and let a, b, c be elements of L such that $a \geq b$ and $a \cup c = b \cup c$, $a \cap c = b \cap c$. Then

$$a = a \cap (a \cup c) = a \cap (b \cup c) = b \cup (a \cap c)$$

$$= b \cup (b \cap c) = b.$$

Conversely suppose that L is any lattice that satisfies the condition of the theorem. Let $a \geq b$. Then we know that $a \cap (b \cup c) \geq b \cup (a \cap c)$. Also

$$(a \cap (b \cup c)) \cap c = a \cap ((b \cup c) \cap c) = a \cap c$$

and

$$a \cap c = (a \cap c) \cap c \leq (b \cup (a \cap c)) \cap c \leq a \cap c$$

---

* See the remarks on the Jordan-Hölder theorem on p. 200.

so that

$$(b \cup (a \cap c)) \cap c = a \cap c.$$

By duality we have

$$(a \cap (b \cup c)) \cup c = b \cup c$$

$$(b \cup (a \cap c)) \cup c = b \cup c.$$

Hence,

$$a \cap (b \cup c) = b \cup (a \cap c)$$

and L is modular.

We establish next an analogue for modular lattices of the second isomorphism theorem for groups, namely,

Theorem 4.   *If a and b are any two elements of a modular lattice, then the intervals $I[a \cup b, a]$ and $I[b, a \cap b]$ are isomorphic.*

Proof.   Let x be in the interval $I[a \cup b, a]$, so that $a \cup b \geq x \geq a$.   Then $b \geq x \cap b \geq a \cap b$ and $x \cap b$ is in the interval $I[b, a \cap b]$.   Similarly, if y is in $I[b, a \cap b]$, then $y \cup a$ is in $I[a \cup b, a]$. We therefore have a mapping $x \rightarrow x \cap b$ of $I[a \cup b, a]$ into $I[b, a \cap b]$ and a mapping $y \rightarrow y \cup a$ of $I[b, a \cap b]$ into $I[a \cup b, a]$. We shall now show that these are inverses of each other so that either one defines a 1–1 correspondence of one of the intervals onto the other.   Let $x \, \varepsilon \, I[a \cup b, a]$.   Then since $x \geq a$,

$$(x \cap b) \cup a = x \cap (a \cup b).$$

Since $x \leq a \cup b$, this gives $(x \cap b) \cup a = x$.   Dually we can prove that if $y \, \varepsilon \, I[b, a \cup b]$, then $(y \cup a) \cap b = y$.   This proves our assertion.   Since our mappings are evidently order preserving they are lattice isomorphisms.

This theorem leads us to introduce a notion of equivalence for intervals that is stronger than isomorphism.   First we define $I[u,v]$ and $I[w,t]$ to be *transposes (similar)* if there exists elements a,b in L such that one of the pairs can be represented as $I[a \cup b, a]$ while the other has the form $I[b, a \cap b]$.   The intervals $I[u,v]$ and $I[w,t]$ are called *projective* if there exists a finite sequence

$$I[u,v] = I[u_1,v_1], \; I[u_2,v_2], \; \cdots, \; I[u_n,v_n] = I[w,t]$$

beginning with $I[u,v]$ and ending with $I[w,t]$ such that consecutive pairs are transposes. It is immediate that the relation that we have defined is an equivalence. Also by Theorem 4 projective intervals are isomorphic.

We observe now that in the lattice of invariant M-subgroups of any M-group $\mathfrak{G}$ projectivity of a pair of intervals $I[\mathfrak{H},\mathfrak{K}]$, $I[\mathfrak{M},\mathfrak{N}]$ implies M-isomorphism of the factor groups $\mathfrak{H}/\mathfrak{K}$, $\mathfrak{M}/\mathfrak{N}$. It suffices to consider a pair of transposed intervals, say, $I[\mathfrak{H}_1 \cup \mathfrak{H}_2, \mathfrak{H}_1]$ and $I[\mathfrak{H}_2, \mathfrak{H}_1 \cap \mathfrak{H}_2]$. For these, the isomorphism of $(\mathfrak{H}_1 \cup \mathfrak{H}_2)/\mathfrak{H}_1$ and $\mathfrak{H}_2/(\mathfrak{H}_1 \cap \mathfrak{H}_2)$ follows directly from the second isomorphism theorem for groups. This remark will enable us to translate some of the lattice theoretic results to results on group isomorphisms.

EXERCISES

1. Show that, if a lattice is not distributive, then it has a sublattice of order 5 whose diagram is either the first or the second on p. 188. Show also that a non-modular lattice contains a sublattice whose diagram is the first on p. 188.

2. Show that the lattice of subgroups of A_4 is not modular.

3. Prove that, if $\mathfrak{G}$ is a group that is generated by two elements a and b such that $a^{p^m} = 1$, $b^{p^r} = 1$, $b^{-1}ab = a^n$ where $n^{p^r} \equiv 1 \pmod{p^m}$, then any two subgroups of $\mathfrak{G}$ commute. Use this to show that the lattice of subgroups of $\mathfrak{G}$ is modular.

4. Show that if a covers $a \cap b$ in a modular lattice L then $a \cup b$ covers b. A lattice that has this property is called *semi-modular*. Verify that the lattice whose diagram is

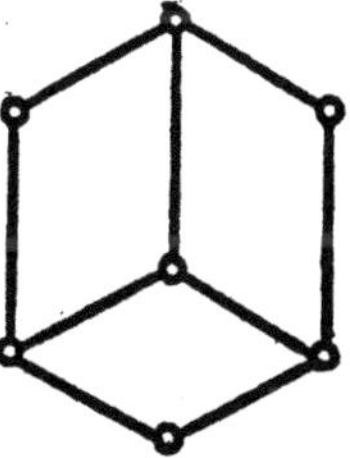

is semi-modular but not modular.

4. Schreier's theorem. The chain conditions.

Let a and b be two elements of a modular lattice satisfying $a \geq b$. We consider now the finite descending chains

$$(2) \qquad a = a_1 \geq a_2 \geq a_3 \geq \cdots \geq a_{n+1} = b$$

connecting a and b. One such chain is called a *refinement* of a second if its terms include all the terms of the other chain. Two chains are said to be *equivalent* if it is possible to set up a 1–1 correspondence between the intervals $I[a_i, a_{i+1}]$ of the two chains such that corresponding intervals are projective. We use these terms in formulating the analogue of Schreier's theorem on groups as follows:

Theorem 5. *Any two finite descending chains connecting the elements a, b $(a \geq b)$ of a modular lattice have equivalent refinements.**

For the proof we require the analogue of Zassenhaus' lemma (third isomorphism theorem). This is the following

Lemma. *Let a_1, a_1', a_2, a_2' be elements of a modular lattice such that $a_1 \geq a_1'$, $a_2 \geq a_2'$. Then the following three intervals*

$$I[(a_1 \cap a_2) \cup a_1', \ (a_1 \cap a_2') \cup a_1'], \quad I[a_1 \cap a_2, (a_1' \cap a_2)$$
$$\cup (a_1 \cap a_2')], \quad I[(a_1 \cap a_2) \cup a_2', \ (a_1' \cap a_2) \cup a_2']$$

are projective.

Proof. Since the second interval is symmetric in the subscripts 1 and 2 and since the third is obtained from the first by interchanging 1 and 2, it suffices to prove that the first and second are projective. Now set

$$a = a_1 \cap a_2, \quad b = (a_1 \cap a_2') \cup a_1'.$$

Then

$$a \cup b = (a_1 \cap a_2) \cup (a_1 \cap a_2') \cup a_1' = (a_1 \cap a_2) \cup a_1'$$

and

$$a \cap b = (a_1 \cap a_2) \cap ((a_1 \cap a_2') \cup a_1')$$
$$= (a_1 \cap a_2') \cup ((a_1 \cap a_2) \cap a_1')$$
$$= (a_1 \cap a_2') \cup (a_1' \cap a_2).$$

This shows that the first interval has the form $I[a \cup b, \ b]$ while the second has the form $I[a, \ a \cap b]$. Hence, these intervals are projective.

Now let

$$(3) \qquad a = a_1 \geq a_2 \geq \cdots \geq a_{s+1} = b$$

$$(4) \qquad a = b_1 \geq b_2 \geq \cdots \geq b_{t+1} = b$$

---

* This form of the theorem is due to Ore.

be two descending chains connecting a and b. As in the group case we introduce the elements

$$a_{ik} = (a_i \cap b_k) \cup a_{i+1}, \quad k = 1, 2, \cdots, t+1$$

$$b_{ki} = (a_i \cap b_k) \cup b_{k+1}, \quad i = 1, 2, \cdots, s+1.$$

Then

$$(5) \quad a = a_{11} \geq a_{12} \geq \cdots \geq a_{1,t+1} = a_{21} \geq a_{22} \geq \cdots \geq a_{2,t+1}$$

$$\geq \cdots \geq \cdots \geq a_{s,t+1} = b$$

$$(6) \quad a = b_{11} \geq b_{12} \geq \cdots \geq b_{1,s+1} = b_{21} \geq b_{22} \geq \cdots \geq b_{2,s+1}$$

$$\geq \cdots \geq \cdots \geq b_{t,s+1} = b$$

are refinements of (3) and (4) respectively. By the lemma $I[a_{ik}, a_{i,k+1}]$ and $I[b_{ki}, b_{k,i+1}]$ are projective. We can therefore use the correspondence $I[a_{ik}, a_{i,k+1}] \rightarrow I[b_{ki}, b_{k,i+1}]$ to prove Theorem 5.

The refinement theorem which we have just proved can be used to derive the Jordan-Hölder theorem for modular lattices. First, we define a *composition chain* connecting a, b, $a > b$ to be a finite sequence

$$a = a_1 > a_2 > a_3 > \cdots > a_{n+1} = b$$

in which each a_i is a cover of a_{i+1}. As in the group case we can establish directly the following Jordan-Hölder theorem:

Theorem 6. *If $a = a_1 > a_2 > \cdots > a_{n+1} = b$ and $a = a_1' > a_2' > \cdots > a_{m+1}' = b$ are two composition chains connecting a and b in a modular lattice L, then $n = m$ and there is a 1–1 correspondence between the intervals $I[a_i,a_{i+1}]$, $I[a_j',a_{j+1}']$ such that corresponding intervals are projective.*

We assume for simplicity now that L contains 0 and 1, and we take $a = 1$, $b = 0$ in the foregoing discussion. Then if there exists a composition chain connecting 1 and 0, L is said to be of *finite length*. The number of intervals in this chain, which is uniquely determined by L, is called the *length* (dimension) of L.

As in the group case (p. 142) we can prove easily that a modular lattice with 0 and 1 is of finite length if and only if the following two chain conditions hold:

Descending chain condition. There exists no infinite properly descending chain, $a_1 > a_2 > a_3 > \cdots$ in L.

Ascending chain condition. There exists no infinite properly ascending chain $a_1 < a_2 < a_3 < \cdots$ in L.

Assume now that L is modular with 0, 1 and that L has finite length. If a is an element of L, the sublattice L_a of elements $x \leq a$ satisfies the same conditions that we have imposed on L. Evidently a is the all element of L_a. We call the length of L_a also the *rank (dimensionality)* $l(a)$ of a. If $a \geq b$, then it is clear that

$$l(a) = l(b) + \text{length } I[a,b].$$

Hence for any a and b in L we have

$$l(a \cup b) = l(a) + \text{length } I[a \cup b, a],$$

$$l(b) = l(a \cap b) + \text{length } I[b, a \cap b].$$

Since $I[a \cup b, a]$ and $I[b, a \cap b]$ are isomorphic, they have equal lengths. Hence

$$l(a \cup b) - l(a) = l(b) - l(a \cap b),$$

or

$$(7) \qquad\qquad l(a \cup b) = l(a) + l(b) - l(a \cap b).$$

This formula is called the *fundamental dimensionality relation* for modular lattices.

The results of this section yield again Schreier's theorem and the Jordan-Hölder theorem for *invariant* M-subgroups of any M-group $\mathfrak{G}$. Isomorphism of the factor groups determined by the intervals of the chains is assured by the projectivity of these intervals. For example, we can easily derive the Jordan-Hölder theorems for chief series and for characteristic series from the lattice results. On the other hand, the lattice theorems that we have given do not apply to ordinary composition series, since the lattice of all subgroups of a group need not be modular. Somewhat more complicated concepts are required to yield the theory of ordinary composition series.[*]

---

[*] See G. Birkhoff, *Lattice Theory*, revised edition (1949), pp. 87–89, and the references given on p. 89.

EXERCISE

1. A subset A of a lattice L is called an *ideal* if (1) $a,b \, \varepsilon \, A$ implies $a \cap b \, \varepsilon \, A$, and (2) $a \, \varepsilon \, A$ and $x \, \varepsilon \, L$ imply $a \cup x \, \varepsilon \, A$. A is a *principal ideal* (a) if A consists of all $x \, \varepsilon \, L$ such that $x \geq a$ for fixed $a \, \varepsilon \, L$.

Prove that L satisfies the descending chain condition if and only if every ideal of L is principal.

Dualize the definition of ideal and the result stated above. (The dual of an ideal is called a *dual ideal*.)

5. Decomposition theory for lattices with ascending chain condition.

We consider next the lattice abstraction of a part of the theory of ideals in Noetherian rings. We assume that L is a modular lattice that satisfies the ascending chain condition. As in the special case of ideals we say that an element $a \, \varepsilon \, L$ is (*intersection* or *meet*) *reducible* if $a = a_1 \cap a_2$ where the $a_i > a$. It is easy to prove (for example, by using the analogue of the principle of divisor induction) that any element of L can be represented as a g.l.b. of a finite number of irreducible elements.

The theory of primary ideals does not carry over to lattices. Here it appears to be necessary to deal exclusively with the concept of irreducibility, and all that we can establish in the way of uniqueness is the comparatively weak result that the number of terms in any two irredundant representations as g.l.b. of irreducible elements is unique. As before, we say that the representation $a = q_1 \cap q_2 \cap \cdots \cap q_m$ is *irredundant* if $q_1 \cap \cdots \cap q_{i-1} \cap q_{i+1} \cap \cdots \cap q_m > a$ for $i = 1, 2, \cdots, m$.

Suppose now that we have any two representations (not necessarily irredundant) of a as

$$(8) \qquad a = q_1 \cap q_2 \cap \cdots \cap q_m = r_1 \cap r_2 \cap \cdots \cap r_n$$

where the q_i and the r_j are irreducible. We propose to show that any q_i can be replaced by a suitable $r_{i'}$, so that we also have

$$a = q_1 \cap \cdots \cap q_{i-1} \cap r_{i'} \cap q_{i+1} \cap \cdots \cap q_m.$$

It suffices to take $i = 1$. We introduce the notation

$$r_j' = r_j \cap q_2 \cap \cdots \cap q_m, \quad j = 1, 2, \cdots, n$$

and note that $a = r_1' \cap r_2' \cap \cdots \cap r_n'$ and $r_i' \leq q_2 \cap q_3 \cap \cdots \cap q_m$. Now, the intervals

$$(9) \quad I[q_2 \cap \cdots \cap q_m, a] = I[q_2 \cap \cdots \cap q_m, q_1 \cap q_2 \cap \cdots \cap q_m]$$

and

$$(10) \quad I[q_1 \cup (q_2 \cap \cdots \cap q_m), q_1]$$

are isomorphic. It follows that, since q_1 is irreducible in (10), a is irreducible in (9). But the decomposition $a = r_1' \cap r_2' \cap \cdots \cap r_n'$ is valid in (9). Hence $a = r_{i'}'$ for a suitable i. This proves the following

Theorem 7. *If $a = q_1 \cap q_2 \cap \cdots \cap q_m = r_1 \cap r_2 \cap \cdots \cap r_n$ are two representations of an element of a modular lattice as g.l.b of irreducible elements, then for each q_i there exists an $r_{i'}$ such that $a = q_1 \cap \cdots \cap q_{i-1} \cap r_{i'} \cap q_{i+1} \cap \cdots \cap q_m$.*

A simple corollary of this result is the uniqueness theorem:

Theorem 8. *The number of terms in any two irredundant representations of an element as g.l.b. of irreducible elements is the same.*

Proof. Applying Theorem 7 we can write

$$(11) \quad a = r_{1'} \cap q_2 \cap \cdots \cap q_m = r_{1'} \cap r_{2'} \cap q_3 \cdots \cap q_m$$
$$= \cdots = r_{1'} \cap r_{2'} \cap \cdots \cap r_{m'}.*$$

Since the decomposition $a = r_1 \cap r_2 \cap \cdots \cap r_n$ is irredundant, all the r_i appear in the last line of (11). Hence $m \geq n$. By symmetry $m = n$.

6. Independence. Suppose that L is a modular lattice with 0 and 1. We call a finite set $a_1, a_2, \cdots, a_n$ of L *(join) independent* if

$$(12) \quad a_i \cap (a_1 \cup \cdots \cup a_{i-1} \cup a_{i+1} \cup \cdots \cup a_n) = 0$$

for $i = 1, 2, \cdots, n$. We have encountered this notion before in the theory of direct products of groups. In this section we shall indicate (mainly in the exercises) how a portion of the theory of direct products can be carried over to lattices. The main result that we shall derive in the text is the following

---

*Note that $2', 3', \cdots$ have a slightly different significance here than in Theorem 7.

Theorem 9. *If the elements a_1, a_2, $\cdots$, a_n are independent, then*

$$(13) \quad (a_1 \cup \cdots \cup a_r \cup a_{r+1} \cup \cdots \cup a_s)$$

$$\cap (a_1 \cup \cdots \cup a_r \cup a_{s+1} \cup \cdots \cup a_t) = a_1 \cup \cdots \cup a_r.$$

Proof. We prove first that

$$(14) \quad (a_1 \cup \cdots \cup a_s) \cap (a_{s+1} \cup \cdots \cup a_n) = 0.$$

This is true by assumption if $s = 1$. Assume now that we have it for $s - 1$. Then

$$(a_1 \cup \cdots \cup a_s) \cap (a_{s+1} \cup \cdots \cup a_n) \leq (a_1 \cup \cdots \cup a_s)$$

$$\cap (a_s \cup a_{s+1} \cup \cdots \cup a_n) = ((a_1 \cup \cdots \cup a_{s-1}) \cap (a_s \cup \cdots \cup a_n))$$

$$\cup a_s = a_s,$$

by modularity and (14) for $s - 1$. It follows that

$$(a_1 \cup \cdots \cup a_s) \cap (a_{s+1} \cup \cdots \cup a_n)$$

$$= (a_1 \cup \cdots \cup a_s) \cap (a_{s+1} \cup \cdots \cup a_n) \cap a_s = 0,$$

since $a_s \cap (a_{s+1} \cup \cdots \cup a_n) = 0$. This establishes (14) for all s. We can now apply the modularity assumption to the left-hand side of (13) to obtain the right-hand side.

A number of useful corollaries can be drawn from (13). Some of these are contained in the following

EXERCISES

1. Show that if a_1, a_2, $\cdots$, a_n is an independent set then any subset is independent. Show also that the elements

$$b_1 = a_1 \cup \cdots \cup a_{r_1}, \quad b_2 = a_{r_1+1} \cup \cdots \cup a_{r_2}, \quad \cdots,$$

$$b_k = a_{r_{k-1}+1} \cup \cdots \cup a_{r_k}$$

where $r_1 < r_2 < \cdots < r_k = n$ are independent.

2. Let a_1, a_2, $\cdots$, a_n be a set of independent elements such that $a_1 \cup a_2 \cup \cdots \cup a_n = 1$. Define

$$b_i = a_1 \cup \cdots \cup a_{i-1} \cup a_{i+1} \cup \cdots \cup a_n.$$

Prove the dual relations:

$$b_i \cup (b_1 \cap \cdots \cap b_{i-1} \cap b_{i+1} \cap \cdots \cap b_n) = 1$$

$$b_1 \cap b_2 \cap \cdots \cap b_n = 0$$

$$a_i = b_1 \cap \cdots \cap b_{i-1} \cap b_{i+1} \cap \cdots \cap b_n.$$

3. Prove that, if the elements $a_1, a_2, \cdots, a_n$ are independent and $(a_1 \cup \cdots \cup a_n) \cap a_{n+1} = 0$, then the elements $a_1, a_2, \cdots, a_{n+1}$ are independent. Prove that the set $a_1, a_2, \cdots, a_n$ is independent if and only if $(a_1 \cup \cdots \cup a_i) \cap a_{i+1} = 0$, for $i = 1, 2, \cdots, n - 1$.

4. Show that, if L satisfies the chain conditions, then the elements $a_1, a_2, \cdots, a_n$ are independent if and only if

$$l(a_1 \cup a_2 \cup \cdots \cup a_n) = l(a_1) + l(a_2) + \cdots + l(a_n).$$

An element a is *(join) decomposable* if $a = a_1 \cup a_2$ where the a_i are independent and $\neq a$. If L satisfies the descending chain condition, then the argument used in the group case (p. 154) shows that any element of L can be represented as l.u.b. of a finite number of independent indecomposable elements.

If $a = b \cup c = b \cup d$ where $b \cap c = 0 = b \cap d$, then the intervals $I[a,b]$ and $I[c,0]$ and the intervals $I[a,b]$ and $I[d,0]$ are transposes. Hence $I[c,0]$ and $I[d,0]$ are projective. We therefore say that the elements c and d are *directly projective* if b exists in L such that

$$b \cup c = b \cup d, \quad b \cap c = b \cap d = 0.$$

This concept is used in the lattice form of the Krull-Schmidt theorem. We state this result without proof as follows:

Theorem. *Let L be a modular lattice with 0 and 1 that satisfies both chain conditions. Suppose that*

$$a = a_1 \cup a_2 \cup \cdots \cup a_m = b_1 \cup b_2 \cup \cdots \cup b_n$$

where the a_i are independent and indecomposable and the b_j are independent and indecomposable. Then $m = n$ and the a_i and b_j can be put in 1–1 correspondence in such a way that corresponding elements are directly projective.

This theorem is due to Kurosch and to Ore.[*] It is immediate that it implies the Krull-Schmidt theorem for groups except for the statement concerning the intermediate decompositions.

[*] See Birkhoff's *Lattice Theory*, rev. ed., p. 94.

7. Complemented modular lattices

Definition 4. *A lattice L with 0 and 1 is said to be* complemented *if for every a in L there exists an a' such that $a \cup a' = 1$,
$a \cap a' = 0$.*

Also if a is any element of a lattice L with 0 and 1, an element
a' such that $a \cup a' = 1$, $a \cap a' = 0$ is called a *complement* of a.
Thus our definition states that a lattice is complemented if and
only if every $a \, \varepsilon \, L$ has a complement. If $b \leq a$, an element b_1
($\leq a$) such that $b \cup b_1 = a$ and $b \cap b_1 = 0$ is called a *complement
of b relative to a*.

The lattice of subsets of a set is complemented. The complement of a subset A is the usual set theoretic complement, that is,
the set A' of elements $a' \notin A$. If all the elements of a finite commutative group have finite prime orders, then the lattice of subgroups of the group is complemented. This will follow from a
criterion that we shall establish presently.

Let L be a complemented modular lattice and let a and b be
any two elements of L such that $b \leq a$. Then there exists an
element b' such that $b \cup b' = 1$, $b \cap b' = 0$. Hence by modularity

$$a = a \cap (b \cup b') = b \cup (a \cap b') = b \cup b_1$$

where $b_1 = a \cap b'$. Since $b \cap b_1 = b \cap a \cap b' = 0$, it is clear
that b_1 is a complement of b relative to a. Thus we see that, if
L is modular and complemented, then relative complements exist
for any $b \leq$ any a in L. Another way of putting this is that for
every a in L the sublattice L_a of elements $\leq a$ is complemented.

The concept of a point plays an important role in the theory of
complemented lattices. An element p of a lattice with 0 is called
a *point* if p is a cover of 0. If L satisfies the descending chain
condition, L contains points; for we can choose an $a_1 > 0$ and,
if a_1 is not a cover of 0, then there exists an a_2 such that $a_1 >
a_2 > 0$. If a_2 is not a point, there exists an a_3 such that $a_1 >
a_2 > a_3 > 0$. By the descending chain condition this process
terminates in a finite number of steps, and it leads to a point in L.

Assume now that L is complemented and that both chain conditions hold. Let p_1 be a point in L and let p_1' be a complement of
p_1. If $p_1' \neq 0$, we can use the descending chain condition on $L_{p_1'}$

to obtain a point $p_2 \leq p_1'$. Since $p_1 \cap p_2 = 0$, $(p_1 \cup p_2) > p_1$. Also $p_1 \cup p_2$ has a complement which, if $\neq 0$, contains a point p_3. Then $(p_1 \cup p_2) \cap p_3 = 0$ and $p_1 \cup p_2 \cup p_3 > p_1 \cup p_2$. Continuing in this way we obtain a sequence of points $p_1, p_2, p_3, \cdots$ such that

$$p_1 < p_1 \cup p_2 < p_1 \cup p_2 \cup p_3 < \cdots.$$

By the ascending chain condition this breaks off after, say, $n(<\infty)$ steps. When this occurs, we know that $p_1 \cup p_2 \cup \cdots \cup p_n$ has 0 as a complement. This means that $1 = p_1 \cup p_2 \cup \cdots \cup p_n$. Thus 1 *is a* l.u.b. *of a finite number of points.* Also we have chosen the p_i so that

$$(p_1 \cup p_2 \cup \cdots \cup p_i) \cap p_{i+1} = 0, \quad i = 1, 2, \cdots, n - 1.$$

Hence, if L is modular, then the p_i are independent (ex. 3, p. 204).

Conversely, suppose that L is any modular lattice with 0 and 1 that has the property that 1 is a l.u.b. of a finite number of points. We shall show that L satisfies the chain conditions and that L is complemented. Let $1 = p_1 \cup p_2 \cup \cdots \cup p_n$ where the p_i are points. We may suppose that the notation is chosen so that $p_1, p_2, \cdots, p_m$ is a maximal independent subset of the set $p_1, \cdots, p_n$. Then we assert that $1 = p_1 \cup p_2 \cup \cdots \cup p_m$; for otherwise there is an $i > m$ such that $p_i \not\leq p_1 \cup p_2 \cup \cdots \cup p_m$. This implies that

$$\bar{p}_i \equiv p_i \cap (p_1 \cup \cdots \cup p_m) < p_i;$$

hence, $\bar{p}_i = 0$. But then $p_1, \cdots, p_m, p_i$ is an independent set contrary to the maximality of m. We therefore have $1 = p_1 \cup p_2 \cup \cdots \cup p_m$. Since the $p_j, j \leq m$, are independent,

$$(p_1 \cup p_2 \cup \cdots \cup p_j) \cap p_{j+1} = 0, \quad j = 1, 2, \cdots, m - 1.$$

Hence the intervals $I[p_1 \cup p_2 \cup \cdots \cup p_{j+1}, p_1 \cup p_2 \cup \cdots \cup p_j]$ and $I[p_{j+1}, 0]$ are transposes, and consequently $p_1 \cup p_2 \cup \cdots \cup p_{j+1}$ is a cover of $p_1 \cup p_2 \cup \cdots \cup p_j$. It follows now that

$$1 = (p_1 \cup \cdots \cup p_m) > (p_1 \cup \cdots \cup p_{m-1}) > \cdots > p_1 > 0$$

is a composition chain for L. The existence of such a chain implies the two chain conditions.

We prove next that L is complemented. Let $1 = p_1 \cup p_2 \cup \cdots \cup p_n$ where the p_i are points. If a is any element of L and $a \neq 1$, we can choose a $p_{i_1} \nleq a$. Then $a \cap p_{i_1} = 0$ and $a_1 = a \cup p_{i_1} > a$. If $a_1 \neq 1$, we can find a p_{i_2} such that $a_1 \cap p_{i_2} = 0$. This process leads to a subset $p_{i_1}, p_{i_2}, \cdots, p_{i_r}$ of the p_j such that

$$a \cap p_{i_1} = 0,$$

$$(a \cup p_{i_1}) \cap p_{i_2} = 0, \quad \cdots, \quad (a \cup p_{i_1} \cup \cdots \cup p_{i_{r-1}}) \cap p_{i_r} = 0,$$

$$a \cup p_{i_1} \cup \cdots \cup p_{i_r} = 1.$$

The first set of equations shows that the set $a, p_{i_1}, \cdots, p_{i_r}$ is independent. Hence $a \cap (p_{i_1} \cup \cdots \cup p_{i_r}) = 0$ so that by the last equation above, $p_{i_1} \cup \cdots \cup p_{i_r}$ is a complement of a.

We summarize our main results in the following

Theorem 10. *If L is a complemented modular lattice that satisfies both chain conditions, then the element 1 of L is a l.u.b. of independent points. Conversely, if L is a modular lattice with 0 and 1 such that 1 is a l.u.b. of a finite number of points, then L is complemented and satisfies both chain conditions.*

A cyclic subgroup of prime order is a point in the lattice $\mathfrak{L}$ of subgroups of a group $\mathfrak{G}$. Hence if $\mathfrak{G}$ is finite and commutative and every element of $\mathfrak{G}$ is of prime order, then $\mathfrak{L}$ satisfies the chain conditions, is modular and 1 in $\mathfrak{L}$ is a l.u.b. of points. We therefore have the proof of the statement made above that $\mathfrak{L}$ is complemented.

EXERCISE

1. Show that for a complemented modular lattice either one of the chain conditions implies the other.

8. Boolean algebras

Definition 5. *A Boolean algebra is a lattice with 0 and 1 that is distributive and complemented.*

The most important example of a Boolean algebra is the lattice of subsets of any set S. More generally any *field of subsets* of S, that is, any collection of subsets which is closed under $\cup$ and $\cap$ and which contains $1 (= S)$ and $0 (= \varnothing)$ and the complement of any set in the collection, is a Boolean algebra.

The following theorem gives the most important elementary properties of complements in any Boolean algebra.

Theorem 11. *The complement a' of any element a of a Boolean algebra B is uniquely determined. The mapping $a \to a'$ is 1–1 of B onto itself; it is of period two ($a'' = a$); and it satisfies the conditions*

$$(15) \qquad (a \cup b)' = a' \cap b', \quad (a \cap b)' = a' \cup b'.$$

Proof. Let a be any element of B and let a' and a_1 be elements such that $a \cup a' = 1$, $a \cap a_1 = 0$. Then

$$a_1 = a_1 \cap 1 = a_1 \cap (a \cup a') = (a_1 \cap a) \cup (a_1 \cap a')$$
$$= a_1 \cap a'.$$

Hence, if, in addition, $a \cup a_1 = 1$, $a \cap a' = 0$, then $a' = a' \cap a_1$. Hence, $a' = a_1$. This proves the uniqueness of the complement. It is now clear that a is the complement of a'; hence, $a'' \equiv (a')' = a$. This proves that the mapping $a \to a'$ is of period two. Consequently it is 1–1 of B onto itself. Now let $a \leq b$. Then $a \cap b' \leq b \cap b' = 0$ so that

$$b' = b' \cap 1 = b' \cap (a \cup a') = (b' \cap a) \cup (b' \cap a')$$
$$= b' \cap a'.$$

Hence $b' \leq a'$. Since $a \to a'$ is 1–1 of B onto itself and is order-inverting the argument used to prove Theorem 1 shows that (15) holds.

Historically, Boolean algebras were the first lattices to be studied. They were introduced by Boole in order to formalize the calculus of propositions. For a long time it was supposed that the type of algebra represented by these systems was of an essentially different character from that involved in the familiar number systems. This is not the case, however. On the contrary, as we shall see, the theory of Boolean algebras is equivalent to the theory of a special class of rings. The proof of this fact is based on the result that any Boolean algebra can be considered as a ring relative to suitably defined compositions.

In order to make a ring out of a Boolean algebra B we introduce the new composition

$$a + b = (a \cap b') \cup (a' \cap b)$$

which is called the *symmetric difference* of a and b. It is immediate that $(a \cap b') \cup (a' \cap b) = (a \cup b) \cap (a \cap b)'$. Thus in the special case of subsets of a set S the symmetric difference $U + V$ is just the totality of elements that belong to U and to V but not to both sets. We shall now show that B is a ring relative to $+$ as addition and $\cap$ as multiplication. From now on we use the customary ring notation ab for $a \cap b$.

Evidently $+$ is commutative. To prove associativity we note first that

$$(a + b)' = (a \cap b) \cup (a' \cap b').$$

Hence,

$$(a + b) + c = \{((a \cap b') \cup (a' \cap b)) \cap c'\}$$
$$\cup \{((a \cap b) \cup (a' \cap b')) \cap c\}$$
$$= (a \cap b' \cap c') \cup (a' \cap b \cap c')$$
$$\cup (a \cap b \cap c) \cup (a' \cap b' \cap c).$$

This is symmetric in a, b and c so that in particular, $(a + b) + c = (c + b) + a$. Commutativity therefore implies the associative law. Evidently,

$$a + 0 = (a \cap 1) \cup (a' \cap 0) = a$$

and

$$a + a = (a \cap a') \cup (a' \cap a) = 0.$$

Hence B is a commutative group relative to $+$.

We know, of course, that $\cdot\ (= \cap)$ is associative. It therefore remains to check the distributive law. This law follows from

$$(a + b)c = ((a \cap b') \cup (a' \cap b)) \cap c$$
$$= (a \cap b' \cap c) \cup (a' \cap b \cap c),$$
$$ac + bc = ((a \cap c) \cap (b \cap c)') \cup ((a \cap c)' \cap (b \cap c))$$
$$= ((a \cap c) \cap (b' \cup c')) \cup ((a' \cup c') \cap (b \cap c))$$
$$= (a \cap c \cap b') \cup (a' \cap b \cap c).$$

Hence $B, +, \cdot$ is a ring.

We note also the following properties of $B,+,\cdot$. The ring is commutative, it has an identity and all of its elements are idempotent. All of these are familiar properties of the composition $\cap$ of any lattice with 1. Also we have seen that every element of B is of order ≤ 2 in its additive group. These statements about a ring are, however, not independent; for, as we now note, $a^2 = a$ for every a in a ring implies $2a = 0$ and $ab = ba$ for every a,b. To prove this we note that

$$a + b + ab + ba = a^2 + b^2 + ab + ba = (a + b)^2 = a + b.$$

Hence

$$(16) \qquad\qquad\qquad ab + ba = 0.$$

If we set $a = b$ in (16) and use the idempotency of a, we obtain $2a = 0$; hence, $a = -a$. Then by (16) $ab = ba$. Thus, the essential facts about $B,+,\cdot$ are that it has an identity and that all of its elements are idempotent. We therefore introduce the following

Definition 6. *A ring is called* Boolean *if all of its elements are idempotent.*

We shall show next that any Boolean ring $\mathfrak{B}$ with an identity defines a Boolean algebra. In order to reverse the process just applied we now define $a \cup b = a + b - ab$ and $a \cap b = ab$. We have seen in Chapter II (p. 56) that $\cup$ (the circle composition) is associative. The other rules in L_1–L_4 are immediate from our assumptions and the commutativity of $\mathfrak{B}$ noted above. Hence $\mathfrak{B}$, $\cup$, $\cap$ is a lattice. This lattice is distributive since

$$(a \cup b) \cap c = (a + b - ab)c = ac + bc - abc$$
$$= ac + bc - acbc = (a \cap c) \cup (b \cap c).$$

Also it is immediate that 1 and 0 are, respectively, the all element and zero element of the lattice and that $a' = 1 - a$ acts as the complement of a. Hence, $\mathfrak{B}$ is a Boolean algebra.

Finally, we note that the two processes that we have applied are inverses of each other. Thus suppose that we begin with a Boolean algebra B, $\cup$, $\cap$. Then we obtain the ring $B,+,\cdot$ where $a + b = (a \cap b') \cup (a' \cap b)$, $ab = a \cap b$. An application of the second process to $B, +, \cdot$ gives the compositions $a \,\overline{\cup}\, b \equiv a + b - ab$ and $a \,\overline{\cap}\, b = ab \equiv a \cap b$. Now $1 - a = 1 + a = (1 \cap a') \cup (1' \cap a) = a'$. Hence

$$a \,\overline{\cup}\, b = a + b - ab = 1 - (1 - a)(1 - b) = (a' \cap b')'$$

$$= a \cup b.$$

Thus the compositions $\overline{\cup}$, $\overline{\cap}$ coincide with the original $\cup$, $\cap$. On the other hand, suppose that we start with a Boolean ring with 1 and we define $a \cup b = a + b - ab$, $a \cap b = ab$ and $a \oplus b = (a \cap b') \cup (a' \cap b)$, $a \odot b = a \cap b = ab$, then $a' = 1 - a$ and

$$a \oplus b = (a \cap (1 - b)) \cup ((1 - a) \cap b)$$

$$= a(1 - b) \cup (1 - a)b$$

$$= (a - ab) \cup (b - ab)$$

$$= a - ab + b - ab - (a - ab)(b - ab)$$

$$= a - ab + b - ab - ab + ab + ab - ab$$

$$= a + b.$$

Hence $\oplus$ coincides with $+$, $\odot$ with $\cdot$. This completes the proof of the following theorem which is due to Stone

Theorem 12. *The following two types of abstract systems are equivalent: Boolean algebra, Boolean ring with identity.*

EXERCISES

1. Show that any Boolean algebra defines a ring relative to the two compositions $a \oplus b = (a \cup b') \cap (a' \cup b)$, $a \odot b = a \cup b$. Show that $a \oplus b = 1 + a + b$, $a \odot b = a + b + ab$ where $+$ and $\cdot$ are as defined in the text.

2. Show that, if e and f are idempotent elements of a ring and $ef = fe$, the ef and $e + f - ef$ are idempotent. Prove that the idempotent elements that belong to the center of any ring with an identity form a Boolean algebra relative to the compositions $e \cup f = e + f - ef$, $e \cap f = ef$.

3. Prove that any ring for which there exists a prime p such that $pa = 0$, $a^p = a$ for every a in the ring is commutative.

INDEX

Adjoint of a matrix, 59
Algebraic element, 94, 183
Algebraic extension of a field, 101
Algebraic integer, 181
Algebraically independent elements, 105
Alternating group, 37
    simplicity of, 139
Anti-homomorphism, 74
Anti-isomorphism, 72
Associated primes of an ideal, 174, 179
Associates, 114
Associative law, 8, 15
    generalized, 20
Automorphism:
    group of, 45
    inner, 46
    of group, 45
    of module, 165
    of ring, 68

Binary composition, 4
    non-associative, 18
Binomial theorem, 52
Boolean algebra, 207
Boolean ring, 210

Cayley's theorem, 28
Center of group, 46
Center of ring, 64
Chain, 188
    composition, 199
    equivalence of, 198
    refinement of, 197

Chain conditions:
    for groups with operators, 142, 153, 154
    for lattices, 200
    for modules, 168
Characteristic of ring, 74
Characteristic subgroup, 130
Circle composition, 56
Closure (in a semigroup), 25
Cofactor of a matrix, 59
Commutative law, 8, 21
Commutator, 132
Commutator group, 132
Complement (in a lattice), 205
Composition:
    binary, 4
    non-associative, 18
    ternary, 18
Conjugate classes, 47
Coset, 37
Cover (in a lattice), 188

Decomposable element (in a lattice), 204
Difference ring, 66
Dimensionality relation (in modular lattices), 200
Direct product:
    complete, 160
    of groups with operators, 144
    of invariant subgroups, 147
Direct sum, 145
Directly projective elements (of a lattice), 204
Distributive law, 8

Division ring, 54
Divisor (factor), 13, 114
   of ideal, 173

Eisenstein's irreducibility criterion, 127
Endomorphism:
   normal, 150
   of group, 45
   of module, 165
   radical of, 155
   ring of, 80
   sum of, 151
Equivalence classes, 5
Equivalence relation, 4
Extension of a field, 100
Extension of a ring, 84
Euler-Fermat theorem, 67
Euler ϕ-function, 34, 67, 121

Factor, *see* Divisor
Factor group, 41, 131
Field, 54, 183
   extension of, 100
   prime, 103
   structure of, 103
Field of fractions, 88
Field of subsets, 207
Fitting's lemma, 155
Fractions, 88

Gauss' lemma, 125
Greatest common divisor, 13, 118
   of ideals, 173
Group, 23
   cyclic, 30
   generators of, 31
   multiplication of, 29
   of automorphisms, 45
   regular realizations of, 29
   simple, 139
   solvable, 139
Group with operators, 128
   decomposable, 152

Group with operators (*Cont.*)
   determined by a ring, 130
   direct product of, 145
   factor group of, 131
   homogeneous, 158
   maximal invariant subgroup, 140
   subgroups of (M-subgroups), 130

Hilbert basis theorem, 171
Holomorph of group, 47
Homomorphism of groups, 41
   fundamental theorem for groups
     with operators, 133
   fundamental theorem of, 44
   kernel of, 43
   natural, 43
   with operators, 131
Homomorphism of lattices, 192
Homomorphism of modules, 165
Homomorphism of rings, 68
   fundamental theorem of, 70
   kernel of, 69

Ideal, 65
   associated prime, 174
   in a lattice, 201
   left, 77
   primary, 174
   prime, 173
   principal, 77
   radical of, 173
   reducible, 175
   regular, 167
   right, 77
Idempotent element, 24
Identity element, 22
   of lattice, 192
Imbedding of commutative integral domain in a field, 87
Imbedding of ring in ring with an
   identity, 84
Independence in lattices, 202
Induction, 7, 9

Integers, 10
  Gaussian, 123
  in quadratic fields, 184, 186
Integral dependence, 181
Integral domain, 53
  Euclidean, 122, 186
  Gaussian, 115, 184
  principal ideal, 121
Integrally closed, 183, 184
Intervals (quotients) in a lattice, 192
  projective, 196
  transpose, 196
Inverse, 22
Irreducible element, 115
Irreducible element of a lattice, 201
Irreducible polynomial, 101
Irreducible (prime) integer, 67
Irredundant intersection:
  of elements of a lattice, 201
  of ideals, 177
Isomorphism:
  of groups, 26
  of lattices, 192
  of modules, 165
  of rings, 68
Isolated components (of an ideal), 180
Isomorphism theorems for groups with operators, 135

Jordan-Hölder theorem, 141
  for lattices, 199

Krull-Schmidt theorem, 156
Kurosch-Ore theorem, 204

Lagrange's theorem, 39
Lattice, 189
  complemented, 205
  complete, 189
  composition series in, 199
  distributive, 193
  modular, 194

Lattice (*Cont.*)
  principle of duality in, 190
  semi-modular, 197
Least common multiple, 14, 120
  of ideals, 173
Leibniz's theorem, 100
Length of element of a Gaussian semi-group, 116
Length of element of a lattice, 199
Linearly ordered set (chain), 188

Mapping, 3
  graph of, 3
  induced by an equivalence relation, 6
  inverse, 4
  inverse image of, 6
  order preserving, 192
  resultant of, 4
Matrix, 56
  adjoint, 59
  cofactor of, 59
  determinant of, 58
  diagonal, 64
  ring, 56
  scalar, 64
  transposed, 72
Maximum condition, 169; *see also* Chain conditions
Minimum condition, 169; *see also* Chain conditions
Möbius function, 120
Module, 162, 163
  annihilator, 165
  cyclic, 166
  difference, 165
  generators of, 166
  modules of a ring, 164
  quotient, 165
  unitary, 167

Newton's identities, 110
Nilpotent element, 55

Order of an element of a group, 32
Order of an element of a module, 165
Order of semi-group, 17

Partially ordered set, 187
Peano's axioms, 7
Permutations, 27
    decomposition into cycles, 34
    even and odd, 36
Poincaré's theorem, 40
Point (in a lattice), 205
Polynomials, 93, 97
    cyclotomic, 127
    homogeneous, 108
    in several elements, 105
    irreducible, 101
    polynomial functions, 111
    primitive, 124
    symmetric, 107
Power series, 95
Powers, 21
Prime element, 14, 116
Projection, 150
    primitive, 158

Quadratic extensions of rational field, 184
Quasi-regular, 55
Quaternions, 60
    norm of, 63
    trace of, 63
Quotient group, *see* Factor group
Quotient in a lattice, 192
Quotient of submodules, 165

Radical of ideal, 173, 175
Realization of a group, 28, 30
Relation, 4
    asymmetry of, 9
    reflexivity of, 5
    symmetry of, 5
    transitivity of, 5

Ring, 49
    additive group of, 50
    Boolean, 210
    commutative, 53
    extension of, 84
    group of units of, 54
    identity of, 53
    multiplications of, 82
    multiplicative semi-group of, 50
    Noetherian, 172
    of formal power series, 95
    of polynomials, 92
    right annihilator of, 82
    simple, 70

Schreier's refinement theorem, 138
    for lattices, 198
Semi-group, 15
    Gaussian, 115
    group of units of, 25
    multiplication table of, 17
    ring, 95
Series:
    characteristic, 143
    chief, 143
    composition, 140, 143, 199
    fully invariant, 130
    normal, 138
Sets, 2
    intersection of, 2
    logical sum of, 2
    product set, 3
    quotient set, 5
Stone's theorem, 211
Subdirect product (of groups), 160
Subfield, 87
Subgroup, 24
    characteristic, 130
    cosets of, 37
    fully invariant, 130
    generated by a subset, 30
    index of, 39
    invariant (normal), 40

Subgroup (*Cont.*)
  left cosets of, 39
  products of subgroups, 76
Sublattice, 191
Submodule, 164
Subring, 61
  division, 63
  generated by a subset, 63
Symmetric difference, 209
Symmetric group, 27

Transcendental element, 93
Transcendental extension of a field, 101
Transformation group, 27
  transitive, 37
Transformations, 4

Transitivity set, 37
Transpositions, 36

Uniqueness of factorization in semigroups, 117
Uniqueness theorems for representation of ideals as intersections of primary ideals, 177
Unit element, *see* Identity element

Vector space, 167

Well-ordering (of natural numbers), 9
Wilson's theorem, 104

Zero divisor, 53